The Ethics of Human Systems

Addressing major social and ecological issues, this book establishes the field of ethics as a fundamental science that cannot be separated from economics and business. In particular, it redefines the nature of good investments which, besides being profitable, have to realize values as opposed to reinforcing the countervalues that are destroying our world.

The book develops a universal morality hypothesis constituted by four sets of universal values-countervalues that correspond to the fundamental laws of ethics. This new understanding of ethics makes human systems the architects that can save our world through doing something for others that satisfies need rather than greed. This book argues that we can realize values through providing something of value to others, that is, human systems can create economic value with impact for good.

The Ethics of Human Systems is an insightful read for academics and practitioners in the fields of systems thinking, business ethics, and impact investing in particular.

Manel Pretel-Wilson is social entrepreneur and has degrees in business studies and philosophy, master's degrees in international relations and sustainability, a PhD in systems science, and senior management experience in innovation and sustainability.

Systems Thinking
Series Editors:
Gerald Midgley University of Hull, U.K.

Systems Thinking theory and practice is gaining ground in the worlds of social policy and management.

The Routledge Systems Thinking series is designed to make this complex subject as easy for busy practitioners and researchers to understand as possible. It provides range of reference books, textbooks and research books on a range of themes in systems thinking, from theoretical introductions to the systems thinking approach and its history, through practical guides to the implementation of systems thinking in the world, through to in depth case studies that are significant for their profound impact.

This series is an essential reference point for anyone looking for innovative ways to effect systemic change, or engaging with complex problems.

Systems Theory and Agile Brand Management
The Educative View of Branding
Jan Lies

Managing Multi-Dimensional Complexities in Organizations
From Problem-Solving to System-Optimization with the Dimensions Surfacing Heuristics
Mehaad A. Tegally

The Ethics of Human Systems
Creating Economic Value with Impact for Good
Manel Pretel-Wilson

For more information about this series, please visit: www.routledge.com/Systems-Thinking/book-series/STHINK

The Ethics of Human Systems

Creating Economic Value with Impact for Good

Manel Pretel-Wilson

Routledge
Taylor & Francis Group
LONDON AND NEW YORK

First published 2026
by Routledge
4 Park Square, Milton Park, Abingdon, Oxon OX14 4RN

and by Routledge
605 Third Avenue, New York, NY 10158

Routledge is an imprint of the Taylor & Francis Group, an informa business

British Library Cataloguing-in-Publication Data
A catalogue record for this book is available from the British Library

ISBN: 978-1-041-09319-0 (hbk)
ISBN: 978-1-041-09320-6 (pbk)
ISBN: 978-1-003-64949-6 (ebk)

DOI: 10.4324/9781003649496

Typeset in Times New Roman
by Newgen Publishing UK

I dedicate this book to my friend Xavier Majoral

Contents

List of Figures ix
List of Tables x
About the Author xi
Foreword by Jed Emerson xii

PART I
The Fundamental Science of Ethics **1**

1 Introduction 3

2 Economic Value and Impact for Good 8

3 The Domain of Ethics 18

4 Human Beings and Human Systems 32

5 Love-Contempt and Values-Countervalues 42

PART II
The Universal Morality Hypothesis **51**

6 The Value of Knowing 53

7 The Value of Life 67

8 The Value of Fellowship 83

9 The Value of Humanity 101

10 Emotions and Feelings 121

Figures

2.1	Relationship between Purpose and Impact for Good	11
2.2	The Two Offspring of the Extractive Economy	13
2.3	Relationship between Profit and Purpose	15
2.4	Economic Value with Impact for Good	17
6.1	The First Fundamental Law of Ethics: Knowing-Ignorance	65
7.1	The Second Fundamental Law of Ethics: Life-Suffering	81
8.1	The Third Fundamental Law of Ethics: Fellowship-Nationalism	99
9.1	The Fourth Fundamental Law of Ethics: Humanity-Misanthropy	119

Tables

3.1 Human World-Hypothesis 22
3.2 Scales of the Human World 23
4.1 World-Hypothesis 33
10.1 Relationship between Emotions and Experiences 125
10.2 Correlation between Values-Countervalues and Emotions 127
10.3 Relationship between Feelings and Experiences 131
10.4 Correlation between Values-Countervalues and Feelings 135
10.5 Correlation between Values-Countervalues and Emotions-Feelings 136

About the Author

Manel Pretel-Wilson has degrees in business studies and philosophy, master's degrees in international relations and sustainability, a PhD in systems science, and senior management experience in innovation and sustainability. After finishing his doctoral degree, he decided to write his first book, *Utopics: The Unification of Human Science* (2020), to advance the field of philosophy as a foundation for science. In 2022, he started a career as a social entrepreneur to create a new breed of non-profit companies with innovative products that satisfy need rather than greed. However, building an ethical economy also requires a new breed of impact investing this book advocates for.

Foreword

The publication of this volume, *The Ethics of Human Systems: Creating Economic Value With Impact for Good*, could not come at a more opportune moment.

While the ultimate implications of global, recently unleashed forces have yet to make themselves fully known, we live in a time when our fundamental understanding of the roles of companies, governments and non-governmental organizations are all under question, if not direct attack. Simultaneously, we are challenged to navigate our way through a time of what has been called the "poly-crisis," with climate, economic, social, and political transition and, in some ways, breakdown combining with a host of related forces moving through our world. These all combine to challenge how we show up and how we invest our resources, both financial and extra-financial, in pursuit of a future world yet to be revealed through the present haze and confusion of change.

As we struggle to advance a coherent, global response to those crises before us, we find ourselves moving toward a period of multipolar power dynamics with the United States independently moving and being moved out of its dominant position as we see other state actors entering the vacuum left as the United States pulls back towards its newly envisioned "America First" posture. In the process, we leave behind international agreements and understandings which have largely governed our planet's intergovernmental, economic and other relations over the past two hundred plus years.

All this leaves us to question what we thought were a host of norms and assumptions regarding how the world is to be structured and, perhaps more importantly, how we are called to show up within the structures of our world. While economic analytics and financial analysis may help us understand what is changing beneath and around us, the interpretation of such analytics is less a question of an objective function than a set of interlocking social constructs which have guided us over past decades. These constructs must now be reassessed, redefined, and renewed if we are to survive not only as a global economic order meeting the needs of its international stakeholders, but as a global social order of which economics is a part.

With current events overtaking our ability to absorb, reflect upon and understand the changes we're moving through, it may seem a luxury to call for a new approach to defining not only the ethics of who we are and how we're called to show up in the world, but who we choose to be and how we are to position ourselves as individuals and communities of interest. While challenging, this is the task we must undertake if we are to promote a renewed vision of sustained, global health and prosperity. As Pretel-Wilson observes, ethics is less a question of focusing upon what is than what should be and as such, is critical to our potential to survive today's challenges in order to capture the possibilities standing before not only this generation but generations to come.

Many incorrectly believe the creation of financial profit to be a fairly straightforward affair, the logical outcome of identifying and then moving to capture market opportunities by creating efficiently managed organizations drawing upon investment capital to resource actions which then advance economic activity which in turn creates value in a given market which then generates profits to be distributed to shareholders. While in a limited sense true, in a greater sense it is also true, as Pretel-Wilson goes on to observe, that profit is merely the byproduct of the pursuit of purpose. The challenge is how best to define and communicate that sense of purpose through which human, social enterprise is infused with natural and other forms of capital. At present, these various aspects of capital may then be mobilized to efficiently – through traditional and innovative capitalist structures – extract capital in the form of profits to an ever-decreasing group of wealth holders. This understanding of capital's purpose as its own replication – the purpose of capital is to create more capital – is the social construct within which most asset owners, managers, entrepreneurs, regulators and politicians (among many others) have operated for over 400 years.

And on strictly economic terms the world is richer today than ever before.

However, it is also profoundly poorer.

The present, dominant framework defining "wealth" is limited and blind to the many extra-financial costs and burdens incurred by global societies as we pursue a limited, finance only approach to the creation of human systems and economics. What is called for is an evolved, more nuanced understanding of economics, one not based on a bifurcated, "do well and then do good" understanding of our world's economics but rather one that is more holistic and blended, grounded not only in an economic understanding of the world but also a more comprehensive approach. Such an approach to understanding our economic order must be grounded within an ethical framework that allows for multiple components of blended value creation.

The following text is built upon a number of foundational notions, including:

- Ethics is a fundamental human science so we cannot continue acting as if it had nothing to do with business because the laws of ethics apply throughout the human world.
- The laws of ethics redefine the nature of good investments which, besides being profitable, have to realize values as opposed to making money that reinforces countervalues.
- Values are realized by human systems pursuing something good that saves our world and countervalues are reinforced by pursuing something bad that leads to its destruction.
- It is possible to create economic value with impact for good as profit can be the consequence of providing something of value to others that realizes values on an ongoing basis.

These and many other considerations of our world's reality ultimately bring us to the compelling conclusion we must operate within an ethical framework that is evidence-based yet promises to result in not only a financially sustainable economic order, but a global world order capable of sustaining not only human but natural life for centuries to come.

Ethics is the framework which infuses our understanding of the world and how we should operate within it. Our various cultural ethical frameworks of the past have helped guide and direct us over past centuries. What is needed is an evolved ethical framework to direct us going forward, helping us go deeper in order to rise higher and advance further over decades to come. The ideas and concepts explored in the following pages offers us one such framework. And is not so much a starting place as continuation of our shared exploration – one upon which our individual and common future depends.

Jed Emerson
Principal
Blended Value Group
Easton, CT

Part I

The Fundamental Science of Ethics

1 Introduction

Let me start with a question, do we live in a better world? Economists tell us that our standards of living or income per-capita levels have been rising since the Industrial Revolution (1820 onward) and worldwide, describing the 1950–1973 as the "golden age", with large disparities between regions, however. The United States overtook the leader, the United Kingdom, in the 1890s; Japan and the "East Asian miracle" caught up with Western Europe in the 1990s; and, more recently, the world economy saw the rise of the dragon, China. This growth has been impressive given the relative stagnation of our standards of living between the years 1000 and 1820. Economists agree that the engine of economic growth is technological progress, so R&D is at the root of our material progress. Without discoveries and inventions we would have remained at subsistence levels and our life expectancy would never have increased above the late twenties, less than the average age of marriage nowadays. So we do live in a better world compared to the past millennium where an infant was expected to live one third of the time of our children. Of course, the well-being of our species has taken a toll on the well-being of our planet since our world has become richer while our planet is becoming poorer.

Our economy has been systematically extracting natural resources from the lithosphere and the biosphere and, consequently, absorbing increasing amounts of talent and capital to create sectors that transform those raw materials into finished products that reach us through distribution channels. That extractive economy is trying to reduce its impact on the planet by being more efficient in its use of natural resources and disposal of waste that can be recycled, as well as trying to reduce its dependence on fossil fuels and on human-made

DOI: 10.4324/9781003649496-2

substances that cannot be broken down by nature. However, at the rate at which we are extracting natural resources and disposing of our waste, the transition to a "circular economy" sounds promising, though not enough to curb the degradation of our planet. Instead, what we need is to restore its biocapacity, which can only be achieved through a *renewable economy without waste*.

Visionaries like the late Ray Anderson already envisioned a future

> when we have cut our last umbilical cord to the mines and the oil wells, when we no longer dump any waste into the landfills or pollution into the air and water. When we no longer take anything from the earth that the earth cannot renew rapidly and naturally.
>
> (Anderson 2009: 5)

The economy is all about creating value for people, satisfying human needs through providing something of value to others. This is the business of business, the pursuit of plenty rather than the attraction of talent and capital to make money for others. Yet, to accommodate the later while forgetting the real problem, there is a movement claiming that companies should "create value for all stakeholders" or, as they like to call it, "generate positive impact". When, in fact, the purpose of business is to satisfy a market need with a product that clients want to buy. Do not get me wrong, I am not saying that companies do not have to save our world. On the contrary, companies have to avoid the destruction of our world. But, let me say it again, the purpose of business is to provide something of value to others. Creating economic value and saving our world are different things as we can create economic value without realizing values. Do not worry if you don't understand that distinction, you are not alone, which is why I am devoting the next chapter to explain the *difference between creating economic value and impact for good.*

This book is about ethics whose scientific object is *the study of the laws that apply to the moral sphere of the human world in relation to human systems.* I say in relation to human systems because it is them, not human beings in isolation, who realize values and reinforce countervalues and, without them, the human world would not embed values. I say this because values and countervalues do not exist in the vacuum without people been moved by them shaping the human world in as much as they care about others or not. This conception of ethics will come to some surprise to those who have always understood it as a branch of philosophy, moral philosophy. Ethics, like all science, is grounded on philosophy and, therefore, it cannot be one of its branches. Instead, it belongs to the domain of the human sciences together with its other relatives, namely, law, economics, and art. And, as a science, ethics has to develop its own hypotheses and my conjecture is that all peoples and civilizations are driven by the same set of ethical laws, that is, by the same values and countervalues. Thus, I have called this scientific hypothesis *the universal morality hypothesis,* which is in marked contrast to the hypothesis that values are relative to a people or civilization and to the hypothesis that some values are mutually incompatible. There is also another possibility, suggested by Scheler ([1913–1916] 1973) and Hartmann ([1926] 2011), that values could exist independently of human beings, but I have already discarded it because it is human systems that shape the human world and, therefore, values would not exist without being realized by people. Lastly, these philosophers also claim that there is a hierarchy of values with higher and lower values, but I also have to dismiss it because it implies that some values are more valuable than others.

Going back to my notes on my first encounter with ethics in high school through a great philosophy teacher, I can see that Àngel Ros gave us a standard definition of the field: "Ethics is the discipline (study) of morality. Ethics is not concerned with what there is, but with what should be. Ethics is different from science because science deals with what there is (the facts)". What is interesting is the contrast between ethics and science, which is akin to the distinction between values and facts, that is, between "what ought to be" and "what is", and this distinction is usually attributed to Hume (1739), though it was only an afterthought.

> In every system of morality, which I have hitherto met with, I have always remarked, that the author proceeds for some time in the ordinary way of reasoning, and establishes the being of a God, or makes observations concerning human affairs; when of a sudden I am surprised to find, that instead of the usual copulations of propositions, *is*, and *is not*, I meet with no proposition that is not connected with an *ought*, or an *ought not*.
>
> (Hume [1739] 2000: 302)

Besides demonstrating that ethics is, indeed, a theoretical-empirical science, we will have to question the identification of ethics with moral obligation, which Kant suggested with his categorical imperative or moral law to distinguish actions dictated by duty from those done for interest. Otherwise we risk confusing the moral sphere with the juridical sphere, a topic I will cover in the third chapter to distinguish the domain of ethics from the domain of the other human sciences, given that the moral sphere is implied by the other spheres but cannot be reduced to any of them. Likewise, ethics cannot be derived from religion because, though religion is an expression of the universal morality hypothesis, the realization of universal values by peoples and civilizations is not relative to any religion. I do not want to downplay the role of religions as they have been fundamental in uniting civilizations by bringing together different nations, but bringing together different civilizations requires a universal morality. Furthermore, demonstrating that ethics is a human science depends on discovering a universal morality hiding in plain sight, one that drives every people and civilization since the dawn of humanity in the universe. In other words, if ethics is a science, this scientific hypothesis has to apply universally. Otherwise, it would be like saying that the laws that apply to our world do not apply to the rest of the universe.

I have mentioned peoples and civilizations and not human beings because all human beings do things together which is what human systems are and what makes up *peoples and civilizations that constitute different levels of the human world.* Contrary to what Hobbes wanted us to believe, the natural state of human beings is being together rather than avoiding each other for fear of losing what others could take away from us. Instead, human beings do things together all the time from time immemorial. Moreover, in Hobbes' "state of nature", there would be no moral sphere of the human world because human beings cannot realize values alone isolated from each other. That state of nature that preceded civil societies or states is not only hypothetical but impossible given that the moral sphere of the human world has been always there preceding every possible political institution. Indeed, the human world was not born with the constitution of the juridical sphere to enforce laws upon its subjects to ensure peace and protect property. In fact, the state should not be confused with a people because states are human systems and peoples are one of the levels of the human world. Do not hesitate if you don't quite understand what human systems are, Chapter 4 clarifies that scientific concept.

Let me finish this introduction by saying that ethics has to be concerned with what humans do rather than with what they ought to do given that it is related to the realization of purposes. However, not all purposes realize values saving our world since humans also reinforce countervalues that are destroying it. The difference between values and countervalues will become clear in the last chapter of Part I of this book. All I can say at this stage is that they are not defined in opposition to each other. Instead, values, unlike countervalues, are realized ideas that save our world whereas the countervalues of those values do otherwise. All values and contervalues are mutually related and are arranged in quaternities, that is, in groups of four values-contervalues. That symmetry is implying laws that explain the moral sphere of the human world, which is a plausible scientific hypothesis suggesting that all peoples and civilizations are moved by the same fundamental laws of ethics. I do not pretend to have discovered every aspect of this scientific hypothesis which will have to be developed further or replaced with a better explanation using the hypothetico-deductive method of science.

References

Anderson, R. (2009). *Confessions of a Radical Industrialist: Profits, People, Purpose – Doing Business by Respecting the Earth.* Toronto: McClelland & Stewart.

Hartmann, N. (2011. *Ética.* Madrid: Ediciones Encuentro.

Hume, D. (2000). A *Treatise of Human Nature.* Oxford: Oxford University Press.

Scheler, M. (1973). *Formalism in Ethics and Non-Formal Ethics of Values: A New Attempt toward the Foundation of an Ethical Personalism.* Evanston: Northwestern University Press.

2 Economic Value and Impact for Good

The confusion between economic value and ethical values is not trivial and runs back to Aristotle's understanding of virtue as a good and happiness as the highest good. Scheler ([1913–1916] 1973) and Hartman ([1926] 2011) classified eudemonism, or the pursuit of a happy life, as an "ethics of goods", which included good fortune or prosperity in addition to being healthy. It is not that Aristotle praised wealth *per se*, but it was an external good that made our life more comfortable. Another ancient morality that contributed to this confusion between goods and ethics was Epicureanism, which advocated for moderate pleasures to make our life less miserable. This hedonism was incorporated by David Hume ([1739] 2000) and Adam Smith ([1759] 1982) by grounding ethics on passions and emotions, and later it was absorbed by utilitarianism when Bentham ([1789] 1907) proposed his principle of utility: "the greatest happiness for the greatest number". Little did he know that his efforts to provide *The Principles of Morals and Legislation* would lead into the domain of economics or the pursuit of plenty. This is why the confusion between economic value and ethical values persists to this day.

Even Scheler ([1913–1916] 1973), the first philosopher to understand the nature of values, would try to accommodate "the agreeable" and "the useful" in his hierarchy of moral values as lower values. Fortunately, his intuition entered the domain of ethics when it recognized the existence of a different type of value which "cannot be 'reduced' to the values of the agreeable and the useful" (Scheler [1913–1916] 1973: 107), namely, the "vital values" or "the *noble*" that are felt by healthy people through external goods that bring them "well-being". Unlike lower value ranks, higher values such as "the noble"

DOI: 10.4324/9781003649496-3

have the person who experiences those higher feelings as its locus. He even found a correspondence between his theory of "communal forms of togetherness" and his hierarchy of values claiming that "the noble" belongs to "the *natural* unit of the life-community" and "the useful" to the "the society", "an *artificial* unit of individuals having *no* original 'living-with-one-another'" (Scheler [1913–1916] 1973: 528), which presumably corresponded to the economic sphere of the human world. Though creative, it was not very useful to distinguish economic value from ethical values. Hartmann ([1926] 2011) also tried to do his best to distinguish "the values of goods", such as "the value of life" and "the value of happiness", from moral values, claiming that goods are not moral values but ground them since they are "materially founded" on non-moral values.

Unfortunately, both moral philosophers failed to distinguish the moral sphere from the economic sphere because they assumed that one rested above the other, but at least they did not confuse the moral sphere with the economic sphere even if that meant postulating a hierarchy of values. The answer to distinguish economic value from ethical values rests on a true understanding of the goods that satisfy our needs. Scientifically speaking, the great economist that can help us out here is Carl Menger (1871) with his theory of goods, which is, essentially, a theory of economic value. I have already discussed his scientific contribution in my first book, showing that his theory of value is not the same as the so-called theory of marginal utility attributed to Jevons (1871) and Walras (1874). A good satisfies a need because it is useful, and the economic value that we attribute to that good comes from the knowledge of how it contributes to our needs. The crucial difference with the fathers of the so-called Marginalist Revolution in economics is that Menger's theory of value is based on knowledge rather than on the subjective pleasure it produces. Indeed, this is not a subjective theory of value, even if the locus of value is the human being who may well be deceived about the contribution of the goods to its needs or about the alleged needs they satisfy.

As I wrote, "Menger's theory of value is about utilizing knowledge for human welfare rather than about maximizing pleasure" (Pretel-Wilson 2020: 125). Now I should replace "human welfare", which he used, with "human needs" so as not to confuse it with the value of wellbeing. Let me continue quoting myself. "Basically, goods are things that are useful because we *know* they satisfy our needs" (Pretel-Wilson 2020: 126). Indeed, his "General Theory of the Good" is very explicit about it, “Things that can be placed in a causal connection with the satisfaction of human needs we term *useful things*” and one of the prerequisites to qualify as *goods* is “The human knowledge of this causal connection” (Menger 2007: 52). If this condition is not met – either because they do not provide the benefits we thought they did or do not satisfy the need we imagined having – then they “may appropriately be called *imaginary* goods” (Menger 2007: 53). Moreover, Menger’s description of the role of knowledge in the development of the economic sphere of the human world could not be clearer.

> As a people attains higher levels of civilization, and as men penetrate more deeply into the true constitution of things and of their own nature, the number of true goods becomes constantly larger, and as can easily be understood, the number of imaginary goods becomes progressively smaller. It is not unimportant evidence of the connection between accurate knowledge and human welfare that the number of so-called imaginary goods is shown by experience to be usually greatest among peoples who are poorest in true goods.
>
> (Menger 2007: 53–54)

Indeed, the creation of economic value comes from buying goods that we are ready to acquire because we believe that they satisfy our human needs, even though we could be wrong in believing so and buying something that does not. And who provides those goods that are of value to us? The human systems that produce, distribute, offer, or give access to something that satisfies our needs, namely, companies. So the economic sphere of the human world is made of makers-sellers and buyers-users of goods and services that produce economic value because they are useful things to satisfy human needs. Of course, in a better world we would not be buying “imaginary goods” that create no economic value rather than being misled by false claims that lead us to acquire things we do not need. So the first thing we should know is whether the products we are buying satisfy our needs or are just sold for the sake of making money. If what is bought is just making someone feel better for whatever reason that is not actually connected with the satisfaction of his needs, then it might be making money for someone without creating any economic value.

However, it is also possible that someone buys a good that satisfies a need, but the makers-sellers of that product do not realize any ethical values or, worse, they may even be perpetuating the destruction of our world though what they provide to people. So providing something of value to someone does not necessary contribute to a better world, despite the economic value that the given industry is creating by satisfying a market need. Conversely, people could be providing something of value to others that realizes values that save our world. However, the economic sphere is not the only sphere of the human world in which human systems can realize values through what they do for others but is certainly one in which humans are reinforcing countervalues not contributing to a better world. Only purposes that realize values have an impact for good (Figure 2.1).

When I say "purpose" I mean the purpose of any organization that can be any kind of human system, such as companies operating in an industry, governmental organizations regardless whether they are local, regional, national, or international, non-governmental organizations dealing with human problems, or educational or healthcare organizations from primary to higher education or tertiary care. Likewise, when I say "impact for good" I mean the impact of any kind of human system that realizes ethical values through their purposes. However, the relationship between purpose and impact for good is not always the case. There are organizations whose purposes do not have an impact for good because they do not realize ethical values through what they do for others on a regular basis. And to the question as to why they do not have an impact for good if they create economic value, my short answer is that they care little or nothing for others, and sometimes it is because they do not have a purpose and continue working *as if* what they do mattered to someone.

Figure 2.1 Relationship between Purpose and Impact for Good.

Unlike companies, most organizations whose purposes have no impact for good do not depend on the profits that result from providing something of value to others. Instead, they rely on other sources of revenue to fund their activity and, though they surely report their expenditure, that investment has no impact on the people who they are supposed to serve. Without a clear purpose, what they do for others, it is difficult or impossible to assess the impact of an organization. However, if we want to live in a better world, we cannot expect companies to do everything that other organizations are not doing but could do for others if they had the resources. Bettering the economic sphere of the human world is not the only way we can save our world. We are all in it together, so no organization that can make a difference can be excluded from counteracting the equation that is destroying our world, regardless of the sphere of the human world in which they operate primarily. Furthermore, many non-economic organizations are acting as accomplices helping companies that are destroying our world, either turning a blind eye on what they are doing or even enabling them to do so. And by doing so, they also have blood on their hands for letting it happen.

I have said enough about the other organizations that can also make a difference to save our world. Now I am going to focus on the original sin. If we want to counter the destruction of our world, we should start by questioning the rhetoric of those who claim to be saving the world and diverting the attention to the root cause threating our own existence as a species, namely, the *extractive economy*. Extractive is a good adjective that captures what this economy is doing to our planet. As a species, we are systematically extracting natural resources from the lithosphere and the biosphere at a rate that is threatening our planet and affecting the wellbeing of most, if not all, species, including ourselves. In other words, our species is precipitating its own extinction. Well, we always blame this on our dependence or addiction to fossil fuels. However, what is actually fuelling the extractive economy does not lie beneath our planet's crust. We have created an economy fuelled by countervalues, the most obvious of which is greed, which fuels the others. Greed is the root cause of the extractive economy and the probable cause of our demise.

To make an analogy with biology that might be helpful. Through natural selection, species have been branching from the tree of life into different directions, leaving behind their common ancestors to adapt better to their environment. In the case of our species, that common ancestor is the extractive economy but, unlike other species that have disappeared, we are trying to leave behind our legacy by changing its course. The branching of the tree of greed, so to speak, has led us into two opposite directions because we are no longer conformable with its genetic roots and want to make the world a better place. However, like other species, we cannot leave behind the source from which the offspring come from. Our extractive ancestors are still there and are carried by the next generation because they have not left behind their legacy. So we will continue extracting talent and capital from alternative investments that could save our world. If you are wondering about the offspring of the extractive economy that I am referring to, I will not keep you in suspense; they are: the "purpose for-profit" and "charitable non-profit" sectors (Figure 2.2).

The first offspring was the "giving back to society" because making money preceded giving some of it back through philanthropy to fund the charitable-non-profit sector. By describing it as such, I do not mean to say that this sector does not have a legitimate role because it is precisely to give back charity to society with the hope that those donations will improve the lives of the poor and the most vulnerable people. It is a good purpose that shows our generosity, indeed. The only question is whether that purpose has a real impact on our societies or whether it is just dealing with the consequences of the destruction of our world and, hence, no matter how much philanthropy we invest in noble causes, it will never solve the underlying problem. I am not opposed to philanthropy, however. Non-profit organizations have to mobilize talent and capital to ensure the satisfactions of everyone's needs, not just the very poor but anybody that is struggling to make ends meet, and nobody should be excluded from having access to education, healthcare, housing, and work.

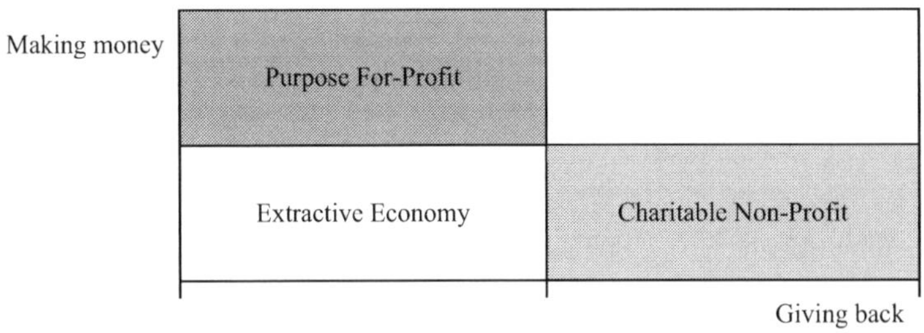

Figure 2.2 The Two Offspring of the Extractive Economy.

The second offspring is the "doing well by doing good" which uses the rhetoric of "pursuing profit with purpose" meaning making money *and* saving the world. This movement is trying to convince us that companies did not have a purpose until then because they were only moved by shareholder return. Accordingly, it is now time for enlightened business leaders to fight the "shareholder primacy" governing companies and to level the playing field to include the concern of other stakeholders to bring forth a "conscious capitalism" that promotes "stakeholder governance". We are told that company boards have to amend their governing documents to include purpose and all the stakeholders they care about to show their commitment to an "inclusive economy". This rhetoric sounds appealing, especially if they change their operating agreement and become so-called benefit corporations to protect the mission of the company. Indeed, it is about our legacy, they stress. But what is our legacy, to consolidate another offspring of the extractive economy? To entice us, they use the word "purpose" profusely as if it was a mantra to convert companies to a good cause that pays off. Yet, they have hijacked a word that is a scientific concept with a clear-cut meaning rather than a marketing slogan that means anything to everyone.

At one stage, when I was writing my master's dissertation, I was also captivated by that idea of blending profit with purpose and questioned the "trade-off mentality" that resulted from John Elkington's (1998) Triple Bottom Line (TBL) and thought that the emerging "Hybrid Corporations" (2012) were going to be "multiple value maximizers" or, to use Jed Emerson's (2003) concept, blended value maximizers. I could not accept the idea that the companies that were going to change the business paradigm, "global value maximizers", were driven by the profit motive alone. As I wrote then, "I think that the main feature that will prevail in the next generation of sustainable enterprises is a drive for multiple value creation, that is, they will enlarge their value proposition to include other forms of value besides financial value" (Pretel-Wilson 2012: 21). With the benefit of hindsight, now I can see why the concept of value cannot be fragmented and compartmentalized but not because the value is "holistic and multidimensional", as I thought at the time. Rather, value is economic value and the creation of economic value implies that companies provide something of value to someone, a product, because it satisfies a human need and, as a consequence, companies derive a profit from that market. In other words, economic value creation implies that the purpose of the company turns a profit

(Figure 2.3) and, the day it does not, it will stop creating economic value because its product-market will not be profitable.

Now that we know the difference between creating economic value and realizing ethical values, we can understand who are the sinners, namely, companies who are not interested in the pursuit of plenty or satisfaction of human needs because they are in business to make money and to cash out as much as they can. These sinners do not care about creating value for others only in making money for themselves. However, the business of business is to create economic value for people and money is made as a consequence of realizing a purpose. Even if they do not want to accept that business of business is not to make money but to satisfy a market need with a product that people acquire, they will have to come to terms with the fact that the pursuit of greed brings with it lots of side effects that are destroying our world.

The pursuit of plenty is not about creating value for investors or workers. These are incidental beneficiaries of what companies do for others. To provide jobs or financial returns is a consequence of providing something useful to others. Companies invest talent and capital to provide something of value to others and, if they are successful, they create jobs and generate profits to pay back investors. The purpose of the company is not to create wealth for its shareholders. Understanding economic value means that companies create value for others and that shareholders are not the beneficiaries of that economic value. This questions the current advocacy of "creating value for all stakeholders", including the environment, as opposed to only creating value for shareholders. The purpose of a company is neither to create shareholder value nor to create value for other stakeholders. So it is

Figure 2.3 Relationship between Profit and Purpose.

about time to stop confusing economic value with profit and social and environmental value with purpose because the purpose of the company is to serve a market need with a product that creates economic value rather than to create value for all stakeholders. Likewise, we cannot say that the purpose of a company is to serve society because its products do not satisfy everyone's needs.

We are getting closer to the understanding of the pursuit of plenty. Companies do not have to pursue everything for everyone. On the contrary, they have to be good at providing something *for someone other than themselves*. If human beings create a company to serve themselves, then they are not in the business of the pursuit of plenty. Furthermore, if they chose to serve the wealthier to make more money when they could offer a product that serves more people, they are neither in that business. The pursuit of plenty implies that companies have to satisfy as many people as they can without becoming unprofitable. If a company tries to serve everyone and becomes unprofitable for doing so, then it has missed its product-market. There are plenty of market needs waiting to be satisfied with products that have not been invented yet. Therefore, the pursuit of plenty has to start with the possible beneficiaries of a product that companies could provide.

The saviours of our world are companies that care about others and thus about making a difference. They contribute to the pursuit of Good through the pursuit of plenty as they shape the moral sphere of our world creating economic value for people. Indeed, they create economic value with impact for good rather than "doing well by doing good", that is, doing good to make money. This is a fallacy because the pursuit of profit is destroying our world. In order to leave behind our true legacy, the extractive economy, the new equation to save our world is not to "make money with positive impact" but to *create economic value with impact for good* (Figure 2.4). Pursuing this new equation is the only possible direction to save our world. I hope business people are taking note rather than following those who do not care about making a difference or are only concerned about their own legacy. Only a true understanding of ethics will help business people who want to make a true difference to save our world before our legacy is its destruction. Any other legacy is pointless if we are not able to save ourselves from ourselves. We can be sinners by commission of ill purposes or by omission for not knowing any better. I will try to remedy the latter with some scientific knowledge that business people are lacking. I do not pretend to know everything

Figure 2.4 Economic Value with Impact for Good.

about ethics so do not expect to find all the answers in a new field of science that has little time to spare to prevent our suffering.

References

Bentham, J. (1907). *An Introduction to the Principles of Morals and Legislation*. Oxford: Clarendon Press.

Elkington, J. (1998). *Cannibals with Forks: The Triple Bottom Line of 21st Century Business*. Gabriola Island, Stony Creek: New Society.

Emerson, J. (2003). The Blended Value Proposition: Integrating Social and Financial Returns. *California Management Review*, 45, 4.

Hartmann, N. (2011). *Ética*. Madrid: Ediciones Encuentro.

Hume, D. (2000). A *Treatise of Human Nature*. Oxford: Oxford University Press.

Jevons, W. S. (1871). *The Theory of Political Economy*. London: Macmillan.

Menger, C. (2007). *Principles of Economics*. Alabama: Ludwig von Mises Institute.

Pretel-Wilson, M. (2012). *Hybrid Corporations: The Next Business Paradigm*. Master's Dissertation. DOI: 10.13140/RG.2.2.33489.56161

Pretel-Wilson, M. (2020). *Utopics: The Unification of Human Science*. Cham: Springer.

Scheler, M. (1973). *Formalism in Ethics and Non-Formal Ethics of Values: A New Attempt toward the Foundation of an Ethical Personalism*. Evanston: Northwestern University Press.

Smith, A. (1982). *The Theory of Moral Sentiments*. Indiana: Liberty Fund.

Walras, L. (1874). *Elements of Theoretical Economics*. Cambridge: Cambridge University Press.

3 The Domain of Ethics

The domain of ethics has been traditionally associated with the *pursuit of happiness* since Aristotle (1910) who argued that "living well" and "doing well" are the same as "being happy". Everyone desires to be happy and everything we pursue in life aim at this goal because happiness is "the highest of all the goods which are the objects of action" (Aristotle 1910). Therefore, the field of ethics has been traditionally concerned with determining *what is good to pursue in life to be happy.* Seneca, the famous Stoic philosopher, wrote a book before he died, *On a Happy Life* (2016), dedicated to his elder sibling with the following opening sentence. "All men, brother Gallio, wish to live happily, but are dull at perceiving exactly what it is that makes life happy" (Seneca 2016: 361). Good and happiness have been inseparable companions since the birth of the field of ethics. However, it has meant not only placing the locus of ethics on one's own happiness, but also making the people that benefit from our actions as means to that end. In contrast, I believe that the object of ethics is not the study of what brings happiness to us, but what saves our world and what destroys it. Technically speaking, the domain of *ethics is the study of the laws that apply to the moral sphere of the human world in relation to human systems*. To become a new field of scientific inquiry, ethics has to redefine what is good and stop identifying it with happiness. *Good is what saves our world and bad is what destroys it.*

We have a long way to go to make ethics a human science, but if we do not start to travel in the right direction we will never get there. Furthermore, the pursuit of our own happiness is what is getting us farther and farther away from any possible milestone, because what has grounded the field of ethics since Aristotle, *selfishness, in conjunction*

DOI: 10.4324/9781003649496-4

with greed, are responsible for the destruction of our world. Instead, the foundation of morality is not self-centered but grounded on the moral sphere of the human world that human systems shape by realizing values and reinforcing countervalues. Hence, the locus of ethics is not the human being but the human world, which is the recipient and custodian of values that human beings realize doing things together for others on a regular basis. Let me be clear about this from the start, morality is not something realized by the best or most excellent people, the magnanimous souls, so to speak, given that values are not realized by heroic actions or great human beings but by humble actions of human beings doing things together for others that save our world every day. That noble selfishness of great people, like Aristotle, has been detrimental to the field of ethics not only because it has missed its object but also because it has been blind to the demiurges or shapers of the moral sphere of the human world, which are not human beings acting alone but *human beings acting together,* that is, human systems, tout court. So the field of ethics has not only to free itself from selfishness but also from thinking that morality is built on spontaneous actions because human beings work together daily to save our world.

Before we distinguish the domain of ethics from the other spheres of the human world, we will have to fight another misconception linked to self-love on which the field of ethics has been founded. If the demiurges of the moral sphere of the human world are human beings acting together for a purpose, then the traditional distinction between ethics and politics that we find in Aristotle (1910) becomes problematic. Indeed, we have to stop assuming that ethics is concerned with individual behaviour and politics with collective behaviour. For the founder of moral philosophy, it was not possible to separate ethics from politics or, more precisely, from Justice, by which he meant "those things which are apt to produce and preserve happiness and its ingredients for the social community" (Aristotle 1910: 126). In other words, it seems that he conceived ethics as part of politics because happy individuals had to live in a good polis, that is, a just city-state, which ensures collective happiness of all individuals by prescribing actions in accordance to virtue. "Justice is in fact perfect Virtue", "because he that has it is able to practice his virtue towards his neighbour and not merely on himself" (Aristotle 1910: 127). Of course, he could not place that "good to others, as it is for another" to ethics because ethics was founded on self-love, so he had to make ethics dependent on politics and argue that Justice "is not part of Virtue but coextensive with

Virtue" (Aristotle 1910: 128). If ethics is not founded on selfishness, we cannot assume that the individual good depends on ethics and the common good on politics because, as I said earlier, good is what saves our world, which implies caring about others rather than pursuing our own happiness and letting politics compensate for what ethics does not cover. Maybe it is about time to acknowledge that politics is closer to a different sphere of the human world, the juridical sphere.

Well, Kant (1785) went as far as reducing the moral sphere to the juridical sphere by identifying ethics with duty without realizing that what is good and what is just are different things. All moral philosophers have tried to establish the relation between morality and law, as they saw that you cannot separate what is good from *what is just*, which explains why Plato and Aristotle tried to establish a relationship between Virtue and Justice. If we go back to Plato's *Republic* (1970), we can see how he tries to convey to his audience that the idea of Justice is the coronation of a perfect city in which every part of the city, that is, every class within the city, behaves according to a different virtue. We have to expect wisdom or prudence from the philosophers that rule the city because they have to behave according to the rational part of the soul; courage or valour from the guardians of the city because that is the best service of the irascible part of the soul; and moderation or temperance from producers and traders and from all the citizens of the city in order to govern the concupiscent part of the soul, so it does not break the harmony of whole. If every part of the city behaves accordingly, then the city will be a city of Justice because every part of the city will do what is good and the whole city what is just. Therefore, Virtue was implied in Justice, which is what Aristotle also tried to establish without reducing Virtue to Justice because both philosophers understood that what is good is related to what is just.

However, even if what is good is implied in what is just it does not mean that what is good is what is just, which is the identification that Kant made when arguing that the person had to act according to the moral law and, thus, obey the categorical imperative dictated by reason. Good actions, he argued, are those that can be prescribed universally. You cannot lie under any circumstance, not even to save your life, for instance. By doing so, he was identifying the sphere of morality with the sphere of duty or obligation, dictating to us what we must do and must not do to act rightly even though it made us miserable because for him dignity was above happiness so ours must be a dignified happiness. We should aim at living with dignity rather than happiness, and only by doing so, we can expect to be happy one day, maybe in another world, because we deserved to be so. So the pursuit of our own happiness was not ruled out, but it would only come from pursuing our own duty because it was a postulate of reason, a rational expectation that could not be proven but was enough to satisfy our hope of receiving that blessing. Furthermore, Kant saw contrary to duty any action that is not dictation by reason. Actions opposed to duty are driven by *interest* or inclination, which he defines as everything that has to do with "happiness, because in this idea all inclinations are summed up" (1785 [2001]: 61). Despite what it might seem at first, he was not questioning selfishness *per se* given that the duty dictated by moral law was grounded on the *respect to oneself* or dignity, so it was still the foundation of morality and happiness was still waiting for us somewhere if we always acted *from duty*. However, he was right about one thing and something that was fundamental for something other than the moral sphere. Indeed, in trying to distinguish good actions, that is, actions acted out of pure duty from non-moral actions motivated by mere inclination, Kant had discovered the defining duality constituting the juridical sphere of the human world, namely, *Duty-Interest*.

We have stumbled upon the juridical sphere, but, so far, we have not said much about the moral sphere of the human world other than saying that it is implied in the juridical sphere but cannot be reduced to it. So both Aristotle and Kant were aware that what is good is related to what is just. And theycould only conceive two possibilities, either Virtue is implied in Justice to the extent of making Justice a Virtue or Justice is implied in Virtue to the extreme of reducing Virtue to Justice. But what is the relationship of the moral sphere to the juridical sphere? Well, a better way to say that what is good is implied in what is just is to say that the moral sphere does not exclude acting out of duty, that is, we can pursue our own duty while realizing values. Indeed, the moral sphere enters the juridical sphere of the human world in the same way as the pursuit of plenty contributes to the pursuit of Good. Another way to see the relationship is that, if the juridical sphere does not exclude the moral sphere, then it is because the latter is included in the former somehow. However, what is included is not one sphere inside the other, because, if that was the case, then we would be reducing the juridical sphere to the moral sphere. Instead, what is implicit is a constituting duality inside another, which is what I mean by saying that the moral sphere is implied in the juridical sphere. By saying that one is implied in the other, we have not said much, because we are still in the dark as to which is the defining duality constituting the moral sphere of the human world. If it does not exclude the duality that Kant found, Duty-Interest, it has to be a fundamental duality that implies that constituting duality but does not reduce it to itself, which means that that duality we are searching for is implied in it but is not coextensive with it. Furthermore, that relationship of *being implied without being coextensive* is what makes each sphere stand out in the constitution of the human world because none is reducible to another, but there is only one sphere that embraces the all the others, the moral sphere, and only one constituting duality that is implied in all the others.

My human-world hypothesis (Table 3.1) suggests that *Love-Contempt* is the fundamental duality defining the moral sphere that is implicit in but not coextensive with the other dualities constituting the other spheres of the human world. Once we have established the relationship between the moral sphere and the juridical sphere, we can just mention the remaining dualities that constitute the economic sphere and the artistic sphere, namely, *Utility-Satisfaction* and *Perfection-Delight*, respectively. All that we need to know at this point is that the juridical duality is implied in the economic duality and that the artistic duality is implied in the economic duality but none is reducible to the other dualities that are implicit in their sphere. Therefore, if each sphere of the human world presupposes the duality of the sphere to which it is related, then Love-Contempt is presupposed in every other sphere of the human world. In other words, the juridical, economic and artistic spheres are not independent from the moral sphere. If so, the relationship between the different spheres of the human world is also suggesting an *order of implication* because, while the moral duality crosses into all the other spheres, the juridical duality is present in the economic sphere and the economic duality in the artistic sphere. That relationship is telling us that the laws that apply to the moral sphere apply equally to the other spheres, but also that the juridical, economic

Table 3.1 Human World-Hypothesis

Human World	*Purposeless Dualities*	*Universal Ideals*
Artistic Sphere	Perfection-Delight	Beauty
Economic Sphere	Utility-Satisfaction	Plenty
Juridical Sphere	Duty-Interest	Justice
Moral Sphere	Love-Contempt	Good

and artistic laws do not cancel out the ethical laws. Likewise, that the juridical laws apply to the economic sphere and the economic laws to the artistic sphere. Conversely, the laws that apply to the economic sphere do not apply to the juridical sphere but neither the artistic laws to the economic sphere. Maybe a better way to understand this relationship is to draw a parallelism with the relationship between the laws that apply to physical, biological, cybernetic (sentient) and human systems. Indeed, the laws of ethics are as fundamental to the human world as the laws of physics to the entire universe.

What about the traditional division between the state and the Church, how does it fit in this scheme of things? It only makes sense if we confuse the spheres of the human world with the human systems that operate within them. Having said that, those well-established human systems are not boundaries within the human world but suggest scales of the human world, such as the national level and the civilizational level in which the human world is structured (Table 3.2). This is not something unique to the human world given that the universe is also structured in scales, given that the Planck constant establishes a boundary between the subatomic and the cosmological scales of the universe. What seems to be unique to the human world is that there is both a subnational and a supracivilizational level as we have subnational and supracivilizational human systems, such as other types of organizations within a nation or country and the United Nations, for instance. Therefore, we should look beyond that traditional Church-state division to avoid confusing the moral sphere with the civilizational boundaries of a religion and the juridical sphere with those of a nation-state because the moral sphere transcends religions and the juridical sphere sovereign states.

If ethics is the study of the moral sphere of the human world in relation to human systems, we will have to start by establishing the domain of ethics, because all that we know so far is that the moral sphere is constituted by the Love-Contempt duality. By the way, before we move forward, I would like to say that discovering the constitution of the

Table 3.2 Scales of the Human World

Human World	Supracivilizational Civilizational National Subnational

human world is the object of one of the domains of philosophy, whose vocation is the discovery of a *human world-hypothesis*, which provides the foundation of the human sciences. The other domain of philosophy, whose vocation is the discovery of a *world-hypothesis*, groundsthe other sciences that study the rest of the universe. I mention this because the development of ethics as a science depended upon the advancement of philosophy, but also because ethics is to the human sciences what physics is to the natural sciences, that is, absolutely fundamental since they all presuppose fundamental laws. However, unlike physics that was grounded on the modern world-hypothesis up until the birth of quantum mechanics (Pretel-Wilson 2020, Annex I), ethics had to wait until the human sciences could ground themselves on a *human world-hypothesis* and make use of the concept of human system. I will devote the next chapter to introducing that scientific concept which can be deduced from a new world-hypothesis I have provided in my first book.

So what can I say about the philosophers that have contributed our understanding of ethics? Well, besides Kant's attempt to ground ethics on the juridical sphere, there have been two approaches to ground moral philosophy. The first attempt was to ground morality on emotions such as pleasure-pain or joy-sadness. Hume (1739), for instance, divided all knowledge into "matters of fact" and "relations of ideas" depending on whether they depended on experimental or logical reasoning. And morality belongs to the domain of matters of fact or existence which, unlike relations of ideas, cannot be demonstrated because "Everything that *is* could *not be*". But moral matters were not a standard empirical object because moral reasoning was based on feeling rather than ordinary perception. If morality was based on experience, then feeling had to correspond to some kind of perception, either "impressions" or "ideas". "Morality", he tells us, "is more properly felt than judg'd of; tho' this feeling or sentiment is commonly so soft and gentle, that we are apt to confound it with an idea" ([1739] 2000: 302). So, if it is not an idea, then it must be some kind of impression. "Here we cannot remain long in suspense, but must pronounce the impression arising from virtue, to be agreeable, and that proceeding from vice to be uneasy" ([1739] 2000: 302). Indeed, the moral distinctions on which the experimental reasoning depends are derived from *pleasure and pain*, "the distinguishing impressions, by which moral good or evil is known, are nothing but *particular* pains or pleasures" (Hume [1739] 2000: 303). And there was a special sense to perceive those moral distinctions, namely, the "moral sense", because they "are not the offspring of reason". "Reason is wholly inactive, and can never be the source of so active a principle as conscience, or a sense of morals" (Hume [1739] 2000: 295).

Likewise, Adam Smith (1759) saw *sympathy* as the cornerstone of ethics. Though to make emotions less subjective, the moral sense gave way to an "impartial spectator" that we all carry within us and allows us to judge others through our feelings."[W]e judge of the propriety or impropriety of the affections of other men, by their concord or dissonance with our own" (Adam Smith [1759] 1982: 16). Sympathy towards other people's feelings is not restricted to emotions such as pity and compassion because Smith gives it a much broader meaning to "denote our fellow-feeling with any passion whatever" (Adam Smith [1759] 1982: 10). Indeed, those "fellow-feelings" that arise within us, either "from the view of a certain emotion in another person" or from finding out what caused them, will "affect the spectator with some degree of a like painful or agreeable emotion" (Adam Smith [1759] 1982: 11). Both moral philosophers coincided that ethics was founded on sentiment or feeling, that is, on emotions or passions, and they both agreed that those moral distinctions or fellow-feelings had to awake a reaction in us, either through our "moral sense" or imagination that allows us to feel a similar emotion. In the case of Smith, however, he assimilated the feeling of other people's grief or joy to the feeling of pleasure and pain that we experience from imagining what they are going through or have gone through. Furthermore, possibly following in Hume's footsteps, who associated "love and hatred" with pleasure and pain, he reduced one to the other. "Love is an agreeable; resentment, a disagreeable passion" (Adam Smith [1759] 1982: 15). In contrast, Spinoza (1677) treated pleasure and pain as joy and sadness experienced by the mind and identified both with love and hate. "Love is the joy accompanied by the idea of an external cause" and "Hatred is the sadness accompanied by the idea of an external cause" (Spinoza [1677] 1970: 130)

The second philosophical approach to ground the field of ethics, attempted by Scheler and Hartmann, was to postulate an *a priori* hierarchy of values that could be apprehended through experience, starting with "sensuous feelings" that could feel lower values, which we saw in the previous chapter. Of course, the problem was that Scheler (1913–1916), who was first to propose such an idea, had to explain the relationship between lower values and higher values. Paradoxically, he claimed that lower values are *founded* on higher values, "the useful", for instance, was founded on "the agreeable" which "is *subordinated* to the value of *healthy* life" (Scheler [1913–1916] 1973: 95). In turn, "the value of the noble", that is, vital values, are founded on "spiritual values" which, in turn, are subordinated to the "value of the holy". The apprehension of "the order of value ranks" depended on the acts of preferring, which enabled the possibility of determining "the height of a value", but the experience of that moral order itself depended on the acts of feeling, such as loving and hating. "Love and hate are acts in which the value-realm accessible to the feeling of a being [...] is either *extended* or *narrowed*" (Scheler [1913–1916] 1973: 261). However, those "value-modalities" were not something intangible because we could experience them in a bearer of values, that is, values could be experienced in something that existed but the being of values itself was "*independent* of their bearer". In order words, Scheler believed that values existed prior to being experienced because they were "*independent* of human organization". Hartmann ([1926] 2011) went even further in claiming that values were independent from human beings arguing that they existed in the sphere of the *ideal being*, as entities with an incomplete being that could become tangible in the *real world* with the help of human beings who experienced them as a demand, as something that "ought to be".

Though interesting and insightful because they saw that emotions were morally relevant, both philosophical approaches failed to establish the field of ethics. Unfortunately, the domain of ethics cannot be established on emotions alone, as it is something we share with sentient beings that do not realize values, and the moral sphere is not something independent from the demiurges of the human world, human systems, which realize values through what they do for others on a regular basis. What all these proposals had in common, however, is that they tried to assimilate love and hate to emotions. Scheler (1913–1916), for instance, postulated "an *emotive apriorism*" of love and hate as the core of ethics, claiming that values did not belong to sensibility or reason because they were independent of our constitution and criticized Kant for "exclud[ing] them from ethics because he cannot assign them to 'reason'" (Scheler [1913–1916] 1973: 4). It was difficult for him to argue that emotions were independent from the sentient constitution of a being that feels values, so he added in a footnote that "Ultimately the apriorism of love and hate is the irreducible foundation of *all* other apriorisms (but we cannot demonstrate this here)" (Scheler [1913–1916] 1973: 64). However, judging from his hierarchy of values, it seems that he was trying to find a parallelism between lower and higher values and the body-soul constitution of a person and placed love and hate at the pinnacle of feelings, "*loving* and *hating* constitute the highest level of our intentional emotive life" (Scheler [1913–1916] 1973: 260). Hartmann ([1926] 2011), will avoid relating feelings to our bodily constitution by relating emotive acts to our spiritual constitution and treating love as a value that opened up for us a "new strata of moral values" with the advent of Christianity, namely, "brotherly love".

Indeed, the last major contribution to the field of moral philosophy, Hartmann's *Ethics* ([1926] 2011), placed love in a special group of values when, in fact, *love is intrinsic to all values and Love-Contempt what constitutes the whole moral sphere.* That duality is neither an emotion nor a value but is *at the root of everything we do in life* given that it informs our choices, but to inform them is not to determine what we choose to pursue in life. We can choose, knowingly or not, to pursue something good or bad, but, equally, decide to do nothing or what others tell us to do. It is up to us whether we want to make a difference to the human world or not. That duality at the root of everything we do is neither pushing us in one direction or another. Instead, Love-Contempt is like Schopenhauer's purposeless Will though without its determinism and without being the essence of the whole universe. Instead, it is *the essence of the moral sphere* which, as we know, spills into the other sphere of the human world, which is why it is at the root of everything we do regardless of what purpose we decide to pursue. Love-Contempt is *the most fundamental duality constituting the human world* because it permeates the whole human world whereas, as we have seen, the other dualities do not reach so far.

Now that we have the duality that grounds the domain of ethics we have to define it better to avoid treating it as a dualism because *Love-Contempt is not the same as love and hate*. As I argued in my first book (Pretel-Wilson 2020), there is a difference between a duality and a dualism. To quote myself, "a dualism is always defined in opposition to something", "what is not one thing necessarily becomes the other". Indeed,

> a duality is different from a polarity in the sense that it does not imply a contrariety. Unlike a polarity, a member of a duality is not defined in opposition to the other member. Another way to see this distinction is to realize that in a polarity the relationship between

> its members is dialectical, one is the antithesis of the other, whereas in a duality, the two members are simultaneous.
>
> (Pretel-Wilson 2020: 53–54)

Being simultaneous members of a duality means that Love-Contempt go together but, of course, one cannot love and have contempt for the same thing at the same time because that would be a contradiction and we cannot pretend to establish a field of science on something that goes against logic. One loves something and has contempt for something else. Furthermore, Love-Contempt are simultaneous members but not opposite partners because, if so, it would be a polarity like love and hate. Instead, contempt includes hate but it cannot be reduced to it because contempt also includes other things besides and beyond hate. Contempt comes from the Latin *contemptus* but that language also provides us with another world, *despectus*, which the English language translates in the verbal form as "to disdain, despise". Spinoza (1677) defines contemptus as "the imagination of anything which touches the mind so little" and despectus as "estimating something too lowly" which is another way of saying that contempt includes disparagement. But one thing is to despise something as being of little worth or to be insensitive or indifferent to it and quite another to hate it. Bernard Shaw (1901) noticed the difference, "The worst sin towards our fellow creatures is not to hate them, but to be indifferent to them: that's the essence of inhumanity" (act 2).

What about love, how do we define it? Well, Plato defined love in the *Symposium* as a desire for something we are lacking and we want to possess because "that of which one is in need and does not have on loves" (2001: 30). A philosopher, for instance, has a desire for wisdom as he is lacking knowledge and wants to be wise, so he is someone that loves what he does not have, "what he does not have and what he himself is not and what he is in need of". However, I do not think that love is about willingness to *possess* something we desire. Why do we have to relate love to having to own something we desire? Rather than wanting to possess something, people that love something want to contribute to the things they love. The problem with Plato's concept of love as Eros is that he associated the things we love with need. Eros was the son of Poros and Penia, of Wealth and Poverty.

> First of all, he is always poor; and he is far from being tender and beautiful, as they may believe, but is tough, squalid, shoeless, and

> homeless, always lying on the ground without a blanket or a bed, sleeping in doorways and along waysides in the open air; he was the nature of his mother [Penia], always dwelling with neediness.
> (Plato 2001: 33)

But he is also resourceful, like his father Poros. "Eros is never either without resources nor wealthy, but is in between wisdom and lack of understanding". Again, this understanding of love implies that we are the ones who benefit from the pursuit of Good, seeking something that satisfies our need. The philosopher, however, is not someone that wants to possess knowledge but someone that loves knowledge and wants to contribute to *our* understanding of the universe, that is, he pursues knowledge for others.

Of course, that which one loves and has contempt for is something we will have to investigate in Chapter 5 devoted to how Love-Contempt is related to values and countervalues. The main purpose of this chapter has been achieved, to establish the domain of ethics as a scientific disciple in its own right that studies the moral sphere of the human world in relation to human systems who shape it. Indeed, the ones who love or have contempt for something are human beings and human systems that have a constitution which is different from the constitution of the human world. That is, humans and the human world do not share the same anatomy. Before we can study the moral sphere and discover what shapes it, we need to introduce the architects of the human world who cannot be confused with their own creation but are responsible for it because it is them who save or destroy our world through what they pursue. Unlike Plato's demiurge who fashioned the cosmos the best he could according to timeless models for which he was not responsible because they were "there" for him to imitate, human systems have had to fashion the human world according to their own ideas. Ideas carried out through purposes which are always informed by the Love-Contempt duality constituting the moral sphere and, due to its fundamental nature, the human world in its entirety. That duality is purposeless but not because the world is purposeless, as Schopenhauer believed, but because it informs every possible purpose that human systems have pursued and will ever pursue, here in our planet Earth or in the rest of the universe. Yes, this fundamental duality of the human world applies to every planet in which human systems shape their own world, that is, in planet Earth *and* elsewhere in the universe.

We can draw a final parallelism between physics and ethics. The laws of physics apply throughout the universe from tiniest subatomic particles to the largest supercluster of galaxies even if some of its fundamental forces, such as gravity, are only perceptible above the Plank constant. Likewise, the laws of ethics apply throughout the human world from the smallest to the biggest human systems in the universe, even if some of its fundamental laws are only seen at certain *scales of the human world,* such as at the national level but also beyond the civilizational level. Lastly, what these two fundamental fields of science have in common is that they share the same number of fundamental laws and that they cannot be unified into a single theory of everything, which means that theoretical physicists will never discover a more general theory to integrate general relativity and quantum mechanics. just like there is no point in searching for a higher synthesis that integrates the laws of ethics.

What this parallelism with physics is telling us is that *the laws of ethics are laws of the human world.* I say this because Hume confused ethics with the study of human nature. "Moral philosophy, or the science of human nature" had to unveil "the secret springs and principles, by which the human mind is actuated in its operations". To repeat it again, ethics studies the moral sphere of the human world in relation to human systems which are the demiurges or shapers of our world. Let me end this chapter with Hume's full quote on the need to transform "moral philosophy" into a science to accomplish what Newton did in "natural philosophy".

> Astronomers had long contented themselves with proving, from the phaenomena, the true motions, order, and magnitude of the heavenly bodies: Till a philosopher, at last, arose, who seems, from the happiest reasoning, to have also determined the laws and forces, by which the revolutions of the planets are governed and directed. The like has been performed with regard to other parts of nature. And there is no reason to despair of equal success in our enquiries.
> ([1748] 1900: 11)

References

Aristotle (1910). *Nicomachean Ethics*, trans. G. H. Lewes. London: George Routledge.

Hartmann, N. (2011). *Ética*. Madrid: Ediciones Encuentro.
Hume, D. (1900). *An Enquiry Concerning Human Understanding*. Chicago: Open Court Publishing.
Hume, D. (2000). A *Treatise of Human Nature.* Oxford: Oxford University Press.
Kant, I. (2001). *Fundamentación de la Metafísica de las Costumbres*. Madrid: Espasa Calpe.
Plato (1970). *Republic*, trans. A.D. Lindsay. London: Heron Books.
Plato (2001). *Symposium*, trans. S. Bernardete. Chicago: University of Chicago Press.
Pretel-Wilson, M. (2020). *Utopics*: *The Unification of Human Science*. Cham: Springer.
Seneca (2016). *Dialogues*, trans. A. Stewart. Standard Ebooks. https://standardebooks.org/ebooks/seneca/dialogues/aubrey-stewart
Shaw, B (1901). *The Devil's Disciple* (First ed.). London: G. Richards.
Smith, A. (1982). *The Theory of Moral Sentiments*. Indiana: Liberty Fund.
Spinoza, B. (1970). *Ethics*, trans, A. Boyle. London: Heron Books.

4 Human Beings and Human Systems

If I could rewrite my first book, I would avoid the mistake of confusing human beings with the human world. That confusion came from not being able to distinguish our constitution from the constitution of our world. The other mistake was to narrow the concept of human beings to utopic beings without realizing that it was better to define them as *creators* since it was much more consistent with the *world-hypothesis* I was introducing (Pretel-Wilson 2020). The last piece of the puzzle was to realize that human systems, which share our same constitution, are human creations. Indeed, besides realizing ideas, human beings create human systems that shape the human world. Indeed, the missing link between the human being and the human world were *the demiurges of the human world.* But I required a scientific understanding of human systems to complement the concept of human beings logically deduced from a new word-hypothesis. In other words, there is a philosophical and a scientific concept of human systems. However, the fact that they are different does not mean that they are distant concepts, since the concept of human systems presupposes the concept of human beings and, thus, it must be logically coherent with it. But before anything of this could be accomplished, philosophy had to provide the concept of system because the universe is made up of systems, not of individuals living next to each other.

The universe or, better, "everything that is (exists)" is constituted somehow. That somehow is the *anatomy of the universe* that grounds all the sciences that study different kinds of systems, namely, physical, biological, sentient or cybernetic, and human systems. Of course, now we know that the human sciences, and the fundamental science of ethics is one of them, study the human world in relation

DOI: 10.4324/9781003649496-5

to human systems, not human beings *per se*, which was the mistake that Hume made when trying to transform "moral philosophy" into "the science of man". Philosophy provides the foundation of science and, therefore, it has to investigate both the anatomy of the universe *and* the anatomy of the human world to make logically possible the advancement of science, from the natural to the human sciences. Well, what I have discovered in pursuing my philosophical investigation is that those two anatomies share something in common as they are both constituted by dualities. We have already introduced the different *purposeless dualities* that constitute the human world at the root of every possible purpose that human beings and human systems decide to pursue. Some of those dualities are more fundamental than others because they are implied in the other spheres of the human world without being reducible to any of them. Similarly, the anatomy of the universe is constituted by different *spatial-temporal dualities* and some are more fundamental than others because they are implied in every type of system that exists in the universe without being reducible to any of them (Table 4.1).

The universe is made up of systems that share the same anatomy and human systems share the same constitution as their creators, namely, the same spatial-temporal duality: *Actuality-Possibility*. That duality arises from pairing together the members of the spatial (inner-outer) and temporal (present-future) dualities intrinsic to human beings: actuality (outer-present)-possibility (inner-future). Furthermore, this duality that defines our constitution implies all the other spatial and temporal dualities intrinsic to other types of systems because we share their constitution plus a different duality that make us unique in the universe. If so, the traditional *body-soul dualism that allegedly constitutes the human being is false*. This old prejudice has persisted because of a lack of understanding of the true constitution of sentient or cybernetic systems, in which the organism cannot be

Table 4.1 World-Hypothesis

Universe	*Spatial Dimension*	*Temporal Dimension*
Human Systems	Inner-Outer	Present-Future
Cybernetic Systems	Central-Peripheral	Past-Present
Biological Systems	Vertical-Horizontal	Origin-Past
Physical Systems	Concentration-Dispersion	Direction-Origin

separated from its environment because they are functionally related and, thus, there is not organism without an environment and, therefore, the brain is not separable from its environment. Unfortunately, that is exactly was what Descartes ([1637] 1913) claimed, the possibility of decoupling the mind from the body, when he defined ourselves as a thinking thing that existed aside an extended body and, thus, postulating a psychophysical parallelism between the mind and the body. That dualism contradicts the constitution of human beings who presuppose the constitution of sentient systems and, therefore, our constitution cannot exclude the spatial (central-peripheral) and temporal (past-present) dualities intrinsic to cybernetic systems because they are implied in the actuality-possibility duality defining us as human beings. Likewise, the spatial (vertical-horizontal) and temporal (origin-past) dualities intrinsic to biological systems are, in turn, implied in cybernetic systems, just like the spatial (concentration-dispersion) and temporal (direction-origin) dualities intrinsic to physical systems are implied in every other system in the universe. In other words, the whole anatomy of the universe, that is, the spatial-temporal constitution of everything that exists, is pressuposed in our human constitution.

I have said that human beings and human systems share the same intrinsic constitution, the Actuality-Possibility duality. However, one thing is to say that human systems the same spatial-temporal anatomy and another to say that *systems share a spatial-temporal structure*. Indeed, this is what it means to be a system because all systems, regardless of the type of system, share a spatial-temporal structure with systems that share the same constitution. That is, systems are beings that do not coexist next to each other but share the space and time with other beings of the same kind. In other words, systems are *entangled beings*. This solves the riddle of the particle-wave duality in quantum mechanics because what we observe is not an aggregate being but a physical system in which beings equally constituted share a spatial-temporal structure. All systems are entangled beings, from entangled particles to entangled civilizations, that is, from the subatomic to the civilizational levels. Physical beings share a spatial-temporal structure, biological beings share an ecological habitat, sentient beings share a cybernetic system made of the organism with its environment, and human beings share a human system with a system of direction. All systems are made of equally constituted beings that share a spatial-temporal structure which is what is meant by saying that they are entangled beings. If so, the individual-society dualism is another false prejudice because human beings are never separate from each other and have never lived in isolation avoiding each other in an hypothetical original state of nature before establishing a civil society where they could coexist, as Hobbes ([1651] 1909) claimed. The natural condition of human beings is being entangled with other human beings sharing a human system with a system of direction. Furthermore, human systems can also be entangled with other humans systems having diferent systems of direction. And, let me repeat it again, a human system is not an "artificial man" as Hobbes argued saying that the state was *like* the individual "though of greater stature and strength than the natural", because human beings and human systems share the same anatomy and thus *are* both equally constituted by the Actuality-Possibility duality intrinsic to humans.

Another thing that I said in my first book is that systems "go towards something from themselves" which is the same as saying that systems are *directed beings*, that is, all systems, from the tiniest to the biggest, have a direction and it comes from themselves given that they are the origin of that direction. In fact, the *Direction-Origin* duality that constitutes the temporal dimension of the physical systems is implied by every other type of system that exists in the universe. Therefore, to be a directed being is something intrinsic to every entangled being and not something exclusive of human beings. If the Direction-Origin duality is implicit in the whole anatomy of the universe, then all systems are the origin of that direction. Indeed, *all systems go towards something from themselves*. This is something similar to what I argued in the case of the Love-Contempt duality constituting the moral sphere that crosses the other spheres of the human world, in the same way as our constitution implies the most fundamental duality intrinsic to physical systems. In other words, the Direction-Origin duality crosses the entire anatomy of the universe without exception. This is the t natural condition of systems because they are all always going towards something. This is implied in the principle of inertia formulated by Descartes ([1641] 1912), but it has always taken for granted something I have questioned in my first book.

To quote myself again,

> rather than thinking that something "coming from something" is more fundamental than something "going towards something", it is the other way around, given that we can only come from something if we have already gone towards something from which it can then be said that we "come from".
>
> (Pretel-Wilson 2020: 46)

It is time to introduce the philosophical concept of human being derived from a new world-hypothesis discovering the anatomy of the universe. To do so, we must start with the duality intrinsic to human beings and human systems, and I am sorry but I cannot avoid the temptation of quoting myself again. However, beware not to confuse human beings with human systems as I did back then.

> What are utopic systems? Well, according to their very nature, these are [directed] systems constituted by the simultaneous actuality-possibility duality. However, we need to specify further this particular type of entangled system. Maybe we can start with the possibility aspect of human systems by saying that, like all genuine dualities, the possible is not something opposed to the actual. On the contrary, these members of the duality are simultaneous because what constitutes the human world today (realized ideas) was something possible (possible ideas) before being actual (actual facts). In addition, the opposite of being possible is being impossible rather than actual. So is it clear that the actual order of the human world is not the opposite of a possible order because before becoming a realized actual order, it was a non-realized possible order or utopia rather than an impossible order or chimera; otherwise it would have never been realized in the human world. Possibility, however, cannot be reduced to what has become the actual order, that is, the outer and the present, as doing so would mean reducing the possible to the actual. Possibility, as we remember, results from crossing the two other members of the spatial and temporal dualities intrinsic to human systems, that is, the inner and the future. Taken together, they are telling something fundamental about utopic systems. Human systems look into the future, not in the sense of anticipating the future as something recurring, as cybernetic systems do all the time, but as something possible, something that could be the case even if it has never been so far. It is not that human systems do not have foresight or that they are not able to anticipate the future based on what has already happened in the past, as this is something that we share with cybernetic systems. But looking into the future is different from looking at the present from the past, the past-present temporal dimension intrinsic to cybernetic systems. Furthermore, this looking into the future as something possible is neither wishful

> thinking nor a prophecy, but something defining our thinking imagination itself.
>
> (Pretel-Wilson 2020: 65)

In this definition of human beings, I had stressed the utopic nature of human beings as beings that are able to imagine something possible, but human systems are equally able to realize something possible, that is, they can imagine something that is possible to realize and, in this sense, human beings are *creators*. Indeed, human beings can create things that they have imagined in some way and realized to some extent, even if they are not identical to what they envisioned originally. Human systems do not create perfect replicas of what they imagine because the process of realizing something possible is not straight forward and many things will not go the way they planned and will have to be rethought until they find the way forward. Yet, the more complicated the endeavour, the more they will need to work together with other people to make it happen. Therefore, although we are impressed with that creative genius of those who are capable of giving birth to something possible that only they could imagine, it is much more likely that we will have to imagine *and* realize together what is possible to make it actual. Although what human systems create may not be something great or revolutionary, what all possible things have in common is that they have to be created by someone. So what seems to define the new concept of human being, that is, the creator, and seems to be its very essence, is that that someone is both a utopic being and a practical being, sometimes more utopic than practical and other times more practical than utopic, but every human being has something of both beings or both beings at the same time. It may also happen that the creator loses some its utopic being and becomes a more practical being, or that the practical being becomes more utopian without ceasing to be practical. In any case, the creator is essentially an *utopico-practical being* because it can never completely lose what defines all human beings.

We have found the essence of human beings as creators, a new concept logically deduced from the new world-hypothesis. Now we need to understand the scientific concept of human system common to all the human sciences and that has to be logically coherent with this new concept of human beings. Human systems, like all systems, are entangled beings that share a spatio-temporal structure and directed beings that go towards something from themselves. What makes them unique is that *human systems are human beings that work together for a purpose*. The concept of working together is something extra that does not come from the concept of being entangled nor of being directed, but can be deduced from the two concepts taken together and goes beyond the general concept of system. Indeed, human systems are the only systems in the universe that *produce something* which comes from the concept of human beings as creators. Human systems produce something for others to satisfy a human demand, for instance, but they can produce all sorts of things, not only economic goods, whatever they can imagine before being produced. The process of creating something imagined includes the process of realizing it together with other human beings to produce it. This is where the concept of purpose completes the concept of human systems. Human systems *produce something for someone, the purpose of the system*. Again, human systems are the only systems in the universe that produce something for someone. And that someone can be themselves or someone else besides themselves for whom they do something *on a regular basis*. This last point is important to define what a purpose is because, strictly speaking, it only applies to human systems and not to human beings. A purpose is something done indefinitely and, to do so, who realizes that purpose may live indefinitely, which does not mean living forever but being alive to realize that purpose continuously.

And the scientific concept of human systems receives the final touch with something we mentioned earlier on since there is always someone that establishes the purpose of the system, namely, the *system of direction* which includes the people who run the system. We also mentioned that human beings realize possible ideas through human systems, which are human creations. Well, it is human beings working together that create human systems and establish their purpose. And, again, human systems go towards something from themselves, because it is them who are the origin of that direction. If so, then the system of direction is not only the origin of the system but also the source of its direction. Indeed, a human system is created by people working together for a purpose who decide its direction through its system of direction or *metasystem*. In functional terms, the system realizes the purpose and the metasystem decides its direction.

Human systems *are* like human beings because they share the same spatial-temporal constitution, that is, they are constituted by the same actuality-possibility duality, and, equally, they also share a spatial-temporal structure with other human beings with whom they are entangled. This is essential to understand the nature of human systems because human beings create human systems that are human beings, not artificial things. Like human beings, human systems share a spatial-temporal structure with other human beings which is the same as saying that they share part of their lives together. Like human beings, human systems come together for a purpose but equally go separate ways when they cannot continue working together for a purpose. Like human beings, human systems can decide with whom they want to share their lives with. Unlike human beings who cannot live forever, human systems can live indefinitely working for a purpose without eventually having to retire. However, human systems are not immortal so they have to find ways to make their purpose relevant to the people who they serve with what they produce on a regular basis.

Well, we are now able to distinguish human beings and human systems from the human world. Human beings and human systems are equally constituted by the same spatial-temporal dualities that define humans and the human world is constituted by purposeless dualities. Therefore, humans and the human world are not equally constituted because they do not share the same anatomy. However, this suggests a gap or a chasm between human beings and the human world because, as we know, it is not human beings *per se* that shape the human world. In other words, we have a world-hypothesis and a human-world hypothesis, the first speaks of a human being who is a creator and the second of a human world that has been shaped by someone. But how does the creator shape the human world? The human world implies that someone has moulded it, an architect, but that architect is not the human being. The missing link between the human being and the human world are the *demiurges or shapers of the human world*, namely, human systems. Indeed, we can cross the chasm between two differently constituted "universes" because human beings can realize ideas in the human world by means of human systems, who are the architects of the human world, given that it is them and not human beings who realize ideas that shape our world.

So what is the actual mechanism by which ideas are realized in the human world? We are already familiar with it. The human world is shaped by ideas brought forth by human systems through their purposes. Purposes are the mechanism by which human systems realize ideas because *purposes are bridges that channel ideas into the human world.* Indeed, it is thanks to their purposes that human systems can realize ideas inthe different spheres that constitute the human world. This also means that without human systems, ideas would come and go with the lives of human beings without leavingtheir traces in the human world. Of course, we could find archaeologists and historians willing to preserve the artefacts and precious works of human beings, like natural historians collecting and cataloguing specimens. But even to preserve them for others to see, we would need libraries and museums created with that purpose in mind, that is, human systems. Ideas can be realized in the human world because the purpose of a humansystem, unlike the purpose of a human being, is indefinite and can even be pursued by other humans systems if they disappear.

It is time to end this chapter with a word about the importance of the concept of human system for the human sciences. In the previous chapter we established the domain of ethics and demarcated the moral sphere from the other spheres of the human world which are studied by other human sciences. Well, now we know that they all share the same scientific concept of human system which this chapter has tried to bring together, starting with an philosophical concept of human being logically deduced from a world-hypothesis and finishing with a scientific concept of human system. This means that there cannot be different concepts of what human systems are since all the human sciences must share the same one. This is what we should expect from any domain of universe studied by a group of sciences. So far we have established the field of ethics, which studies the moral sphere of the human world its relation to human systems, and in this chapter we have introduced the concept of human system for the group of sciences that study the laws that apply to the different spheres of the human world, namely, ethics, law, economics and art. These are the group of human sciences that should be distinguished from the cybernetic sciences, namely, neurophysiology, psychiatry, psychology, andethology, which study organisms with their environment. The concept of human system presupposes the functional relationship between an organism and its environment found in cybernetic systems (Pretel-Wilson, 2020: Annex II) in the same way that cybernetic systems pressupose biological systems that have evolved through natural selection.

References

Descartes, R. (1912). *Meditations on First Philosophy*. London: J.M. Dent.

Descartes, R. (1913). *Discourse on the Method*. Chicago: Open Court Publishing.

Hobbes, T. (1909). *Leviathan or The Matter, Forme and Power of a Commonwealth Ecclesiasticall and Civil*. Oxford: Clarendon Press.

Pretel-Wilson, M. (2020). *Utopics: The Unification of Human Science*. Cham: Springer.

5 Love-Contempt and Values-Countervalues

It is time to start investigating the moral sphere of the human world which, as we know, is constituted by the Love-Contempt duality. It might already have occurred to the reader that values are possible ideas that human systems realize in the human world. And given the pervasiveness of that fundamental duality, he might even start thinking that these possible ideas are not just realized in the moral sphere because it is implied in all the other spheres of the human world and, therefore, they have to be related somehow to other possible ideas realized in them. If so, he might rightly conclude that values come together with other possible ideas shaping the other spheres of the human world. Indeed, the moral sphere is so pervasive that its constituting duality and the values that shape that sphere cannot be disconnected from what is going on throughout the human world. And the mechanism by which human systems realize ideas are purposes because, as said in the previous chapter, purposes bridge the gap between our constitution and the constitution of the human world, but they also link the different spheres of the human world. In other words, the realization of possible ideas in the other spheres through purposes also realizes values in the moral sphere.

Now we are able to understand better why I have insisted that human systems realize values *through what they do for others on a regular basis*, that is, through the purposes that human systems pursue. So we can state it clearly: *human systems realize values through what they pursue*. But human systems not only realize values through what they pursue because, as it has become clear, they can also realize other possible ideas, and the essential relationship between values and purposes holds for all possible ideas that shape the human world. However, the

DOI: 10.4324/9781003649496-6

essential relationship between purposes and values is so fundamental that it is always there even if human systems realize other possible ideas. So fundamental that some moral philosophers, like Nicolai Hartmann, have confused the pursuit of purposes with the pursuit of values, as if human beings pursue values with their purposes. Let me insist on this point, to say that human systems realize values through what they pursue is not the same as saying that human systems pursue values because they realize values pursuing purposes. This fundamental relationship means that the realization of values cannot be separated from the realization of purposes and that we do not pursue values but purposes. Another way of putting it is to say that we have already chosen to pursue a value when we chose to pursue a purpose that realizes a value.

Let me mention some examples. Human systems can pursue the passing of a new law within a state that, if enacted, will help realize a value. Thus, the realization of that possible idea in the juridical sphere will contribute to realize a value in the moral sphere. We only have to think in the laws that human systems have passed to make our societies more equal by outlawing different types of discrimination. Likewise, human systems could realize a possible idea in the economic sphere, for instance, a *non-profit economy* to satisfy human needs rather than one that satisfies human needs to make a profit. In so doing, given that values are mutually related, that possible idea could embed multiple values in the moral sphere. Another example of a possible idea that could be realized in the economic sphere would be a *renewable economy* which, besides pursuing the satisfaction of human needs, could prevent the ecological holocaust we are heading towards. The same applies to possible ideas that could be realized by human systems in the artistic sphere, such as not making art a luxury product only accessible to a minority of wealthy people who can afford to enjoy it. Indeed, the realization of possible ideas in the different spheres of the human world contributes to the realization of values in the moral sphere.

But what is exactly realized in the moral sphere with the pursuit of purposes? Yes, values. But how do values differ from other possible ideas that are also realized in the human world? Well, for starters, *values resemble scientific hypotheses* as much as they have withstood the test of the hypothetico-deductive method because they are possible ideas waiting to be discovered. The difference with other scientific hypotheses is that *values are ideas that have to be realized to be discovered*. But this is no exception to the laws that apply to the universe because the human world is realized by human systems, so any scientific hypothesis that we discover will show up in our world. Furthermore, the laws that we discover in the human world using the scientific method apply to every possible human world. Someone might think that not all realized ideas are scientific laws because some come and go, whereas scientific laws do not. Values are scientific hypotheses that differ from the other possible ideas because they are always there. Maybe the following comparison will help clarify this point. Unlike other possible ideas realized in the human world, we do not have to imagine values before we realize them, and this is why the pursuit of values cannot be confused with the pursuit of purposes. Instead, it is scientists that have to imagine them as a possible hypothesis that might apply to the human world prior to discovering them as ethical laws, which is what all scientists do before discovering scientific hypotheses that apply to the universe.

Before we specify the concept of value as a type of realized ideas, we will have to distinguish it from other concepts of value that have been put forward by the last moral philosophers, Scheler ([1913–1916] 1973) and Hartmann ([1926] 2011), who were also the first philosophers to refer to values as something to be realized. Both associated values to a special type of experience, an emotional experience in which the subject apprehends something that is *not sensible* but in a sensible object, that is, values have to be *felt* in something existing that we experience. And both thought that values have a *being* independent of us, that is, values exist objectively independently of any actual human experiencing them in something real. Values are moral facts of a special sort that cannot be reduced neither to a sensible experience nor to an intellectual experience because feeling a value is not like perceiving a colour or apprehending an idea. Scheler (1913–1916) will tell us that "'Moral facts' therefore do not lie in "inner perception"" or "in the realm of ideal meaning *alone*" (Scheler [1913–1916] 1973:165). Philosophy has fallen prey to assimilating everything either to "sensibility" or to "reason" and values do not fit this scheme. "Moral facts, as opposed to the sphere of meanings, are *facts of material intuition*; an intuition that, of course, is *not* sensible" (Scheler [1913–1916] 1973: 166). He also denies that values are "human inventions" or "conventions". Indeed, "there *are* autonomous moral values" and "these are somehow comprehended in a concrete case" (Scheler [1913–1916] 1973: 176). This will open the door to conceiving values as something prior to what exists, something that cannot be reduced to "a factual state of affairs" because they *are* even if they do not exist yet. "Because values not only span existing and non-existing things but also concern the *transition* from non-existence (of a value) to (its) existence" (Scheler [1913–1916] 1973: 185). In other words, values can be realized in something existing without "being-real".

Of course, this will lead Hartmann to conceive values as belonging to the realm of the *ideal being* and to talk about "values realized in the world" by the moral person who experiences them as something that *ought* to be fulfilled in the *real world*. However, both philosophers identified the realization of values with the execution of moral actions in the real world, but they could not conceive values as being realized in the human world. To reach that understanding, it was necessarily a concept of human being as a creator that can realize possible ideas in the moral sphere through the pursuit of purposes. Furthermore, the realization of values cannot be reduced to individual actions. Values are related to purposes, that is, to what human systems do on a regular basis, not only on special occasions in which they have the possibility of acting morally and feel good about it when most of the time they do otherwise. Values are realized in the human world thanks to human systems through what they do for others regularly. Having said this, Hartmann did associate values with purposes, as something intentionally pursued, and Scheler established a correspondence between his hierarchy of values and pure types of social units, "the collective person is directed in some way towards all types of values, and it must possess a peculiar *consciousness* of them and have *consideration* for them" (Scheler [1913–1916] 1973: 545). Indeed, they both conceived values as something we are aware of in the same way as we are aware of pursuing a goal. But, unlike Scheler, who distinguished values from "all norms, imperatives, demands, etc." founded on "the *being of values*" (Scheler [1913–1916] 1973: 187), for Hartmann, values were "moral imperatives", probably influenced by Kant.

But a value is not a duty, something we must fulfil because it is dictated by our awareness of something we ought to pursue. I would say we do not apprehend values in an ideal order of intelligible things and then decide to realize them in the real world because we cannot contemplate them without being compelled to realize them, as Hartmann claimed. Instead, values are possible ideas realized by human systems without necessarily being aware of which ones they are realized through the pursuit of purposes of which they are fully aware. Values are something humans realize but not so much by being aware of them and then deciding to pursue them because it is the right thing to do, but because, as I said earlier, *we have already chosen to pursue a value when we decide to pursue a purpose that realizes a value.* So it does not make sense to think that we are confronted with a choice between pursuing values of different rank and deciding between them. The choice of values is implicit in every choice of purposes even if we are only aware of our choice of purposes. Indeed, humans chose values every time that they pursue purposes. And at the root of that choice there is the fundamental purposeless duality that informs every possible purpose someone might pursue. However, Love-Contempt can never impose or dictate what someone ought to pursue because that choice is driven by what they love or has contempt for. *That something that someone chooses through what they pursue is either something that saves our world, namely, values, or something that does not, namely, countervalues.* I have introduced the concept of value in relation to the concept of countervalue because there is no value without a countervalue. Furthermore, I have avoided defining values and countervalues in a way that could lead to subjectivist interpretations.

What applies to dualities, which are not dualisms, applies equally to values and countervalues given that they are not concepts defined in opposition to each other, that is, a countervalue is not the antithesis of a value. A value is a possible idea whose realization saves the human world, but a countervalue is not a realized idea and its reinforcement does otherwise. So if a countervalue is not a value, what is it then? Countervalues are something we find always and everywhere in the human world and, hence, do not have to be realized to exist because they are already there. However, the fact that countervalues are there also means that there is something else because values and countervalues are defined in relation to each other, so one cannot be there without the other. In short, *values are realized and countervalues are reinforced in the human world through what human systems pursue on a regular basis.* Of course, by defining values and countervalues in non-subjectivist terms, we are also defining *what is good and what is bad* in non-relativistic terms. If what a human system pursues realizes values, then it is pursuing something good, and, consequently, if what it pursues reinforces countervalues, then it is pursuing something bad. Maybe some people might think that they are pursuing something good when they are in fact pursuing something bad.

There is another implication that might seem less obvious. Human beings and human systems are neither good nor bad by nature, it is what they realize or reinforce in the human world that is good or bad. Indeed, evilness is not related to our constitution, as some philosophers have claimed, but to *what* shapes the human world. The humans have always been seen as the origin of evil even though we are, in fact, the origin of good because countervalues have always existed and will always exist in the human world and, hence, strictly speaking, what is *original* in the human world is not evilness but goodness which does not exist until humans realize values. So what is good or bad is what humans pursue depending on whether it realizes values or reinforces countervalues in the human world.

Moreover, what shapes the human world is not disconnected from the rest of the universe, given that what humans pursue can help or harm others beyond themselves and by harming other non-human systems they are also harming themselves. The human world is not an isolated world in which purposes that benefit some humans are good because they do so, as human systems could be doing something for someone that could be destroying themselves. This is already happening with the ecological holocaust resulting from the monsters we have created shaping the economic spheres of the human world, monsters that are eating up the biocapacity of our planet. So to say that what is good and evil is related to what shapes the human world is not to say that it is unrelated to the rest of the universe. I say this because, since Plato, moral philosophers, with the exception of Schopenhauer, did not have much concern for what happens to the rest of the universe. Aristotle, for instance, understood ethics as part of politics, so it was part of being a good citizen within a political community. In contrast, Schopenhauer was concerned with our treatment of animals and even criticised Spinoza for his lack of compassion towards them. Kant showed some respect for animals, but only to those that have served well their masters. Likewise, for Scheler and Hartmann, all values were referred to the person and the community of people who felt and pursued those values, but it never occurred to them that the realization of values and reinforcement of countervalues in our world includes the rest of the universe.

We are nearly done with the concepts of value and countervalue, the only thing that we are missing is to relate them to love and contempt. We have to explain the relationship between contempt and countervalues from the relationship between love and values given that contervalues are related to values. It is clear that only *people who pursue something that realizes a value love what they do*, and, therefore, love pursuing that purpose. Conversely, people who pursue things that reinforce countervalues cannot love what they do and, therefore, have to be driven by something else. So let me be clear, people who realize values through what they do for others are moved by love, whereas people who reinforce contervalues are moved by something other than love, namely, contempt. Again, contempt is not necessarily hate because people can moved by indifference given that values and countervalues are not opposite concepts. Moreover, if people love what they do for others, they will be afraid of losing it, whereas people that do not love what they do will not mind doing something else. Indeed, there is true commitment to purposes that realize values that cannot be found in purposes that reinforce countervalues. It seems that people who are driven by contempt can detach themselves from what they do and pursue something else because they have no genuine commitment to it. All this suggests that *people who realize values are moved by love whereas people who reinforce countervalues are not.*

It seems that, contrary to what Scheler and Hartmann argued, we have a new concept of value which is not something ideal that belongs to an a priori realm of being independent from our experience waiting to be experienced by someone who is compelled to realize them once they have contemplated the hierarchy of values. On the contrary, *values can only exist in the human world created by human systems and nowhere else.* Values are possible ideas that have been realized by human systems in the moral sphere of the human world by means of pursuing purposes, not by means of making them the goal of our actions. Values and countervalues are not goals to be pursued because they are already implicit in every possible pursuit of purposes. People might not be aware of them, indeed, but the pursuit of purposes is a deliberate choice. However, people might choose not to recognize that pursuing something that reinforces a countervalue is something bad or they might think that pursuing something else in parallel that appears to compensate for what they do might be enough to feel that they are doing something good. Yet, only pursuing something that saves our world on a regular basis is something good and pursuing something that does not on a regular basis is something bad. In other words, no amount of philanthropy will ever compensate for the amount of misanthropy reinforced through what human systems do on a regular basis. We cannot offset the pursuit of something bad with the pursuit of something good because the only way to counteract a contervalue is by realizing the value related to it.

Now that we know what values and countervalues are and that they are related to love and contempt, we have one last thing to do before we can start unveiling all the possible values-contervalues that move all peoples and civilizations in the universe. I have said that values and countervalues are not defined in opposition to each other but in relation to each other. There is no value without a countervalue, but one is not the antithesis of the other. A countervalue is always a countervalue of a value, which means that every countervalue is related to a value. But how is that relationship possible? Well, values and countervalues are related to each other because *people who are moved by the love of a value have contempt for its countervalue and people who are moved by a countervalue have contempt for the value to which it is related*. Another way to put it is to say that people who are moved by a countervalue do not love the value to which it is related as opposed to people who are moved by a value that do love that value. So people who are not moved by the love of a value are moved by the contempt for a value. But that relationship between values and countervalues is reciprocal. People who love a value cannot help having contempt for its countervalue in the same way as people who do not love a value cannot help having contempt for it. That contempt for its related countervalue that people who have a value have comes from the belief that they have to defend it to counteract its countervalue. In contrast, the contempt for a value that people have comes from the false belief that no value can counteract its countervalue. Maybe it is because some people cannot accept the state of our world knowing that values can counteract their countervalues while others are happy with it because they benefit from reinforcing them.

We are now prepared to enter the second part of this book. Given that ethics studies the moral sphere of the human world in relation to human systems and that it is a theoretical-empirical science, it must discover the values-countervalues that constitute the *universal morality hypothesis*. This is a conjecture that has to be tested using the hypothetico-deductive method but not by finding a single observation that disproves the "theory" because the universal morality hypothesis is not constituted by an absolute value. Instead, as I have said repeatedly, values are possible ideas, in the plural, realized by human systems who shape our world, the demiurges of the human world. By implication, this scientific hypothesis cannot be constituted by a single value because human systems realize many different values and reinforce equally different contervalues. Therefore, the moral sphere of the human world is necessarily shaped by a plurality of values and countervalues because there is no value without its countervalue. But how many different values-contervalues can we discover that shape the moral sphere of the human world? My conjecture is that there are four groups of values-countervalues that constitute the universal morality hypothesis and each group is constituted by an equal number of values-contervalues. To find another parallelism with physics, those four groups of values-contervalues cannot be unified into a single "theory of everything" like the four fundamental forces of physics. Indeed, those fundamental values-countervalues are the *four fundamental laws of ethics* which cannot be unified into a more general theory that integrates them further. So let me devote the next part of the book to propose a plausible conjecture for what I believe to be the four fundamental "forces" of ethics that move all peoples and civilizations that exist in the universe.

References

Hartmann, N. (2011). *Ética*. Madrid: Ediciones Encuentro.

Scheler, M. (1973). *Formalism in Ethics and Non-Formal Ethics of Values: A New Attempt toward the Foundation of an Ethical Personalism*. Evanston: Northwestern University Press.

Part II

The Universal Morality Hypothesis

6 The Value of Knowing

When discovering the values that human systems realize in the human world, it is too tempting to derive a value from a countervalue assuming that there is a dialectical relationship between them. This is tempting because the relationship that we have established between a value and a countervalue shows that people who love a value necessarily have contempt for its contervalue. But what holds that reciprocal relationship together is that people either love or have contempt for a value and, if so, they either have contempt for its countervalue or not. Another temptation is to infer one of the values in the group from a countervalue and make it *the* fundamental value, for instance, the value of knowledge from ignorance. This has happened in the history of philosophy and has resulted in treating knowledge as a sphere of its own different from values. However, if values are possible ideas realized in the human world, it makes no sense to distinguish ideas from values. Furthermore, taking the love of wisdom as supreme can miss the other values. In fact, the love of wisdom, which got philosophy started with the Greeks, became identified with the pursuit of knowledge and the discovery of truth, which eventually gave birth to modern science in the 17th century.

Today, the divorce between the pursuit of knowledge and ethics has even got its own prejudice, *the fact-value distinction*, which came from a remark that Hume made when realizing that most of the systems of ethics he had studied derived an *is* from an *ought*. But we do not have to go that far in the history of moral philosophy to see that ethics was removed from the pursuit of knowledge as soon as it was related to something practical rather than theoretical, *prudence rather than episteme*. We cannot blame Plato for that traditional

DOI: 10.4324/9781003649496-8

division in philosophy because in the *Republic* he seems to equate both with wisdom. It was Aristotle (1910), instead, who introduced the classical theoretical-practical division with his division of the soul between the scientific part that deals with matters that do not admit of being different from what they *are* and the calculating part "with which we contemplate those that admit it". Science and philosophy would belong to "theoretical thinking" and ethics and politics to "practical thinking"; and added a further division between matters that belong to the "domain of making", that is, craft making or art, "which has as its object to bring something into existence", and those that belong to the "domain of acting" of which prudence is practical wisdom. To distinguish prudence from knowledge as such, that is, episteme, he wrote.

> BUT Practical Wisdom is employed upon human matters, and such as are objects of deliberation (for we say, that to deliberate well is most peculiarly the work of the man who possesses this Wisdom), and no man deliberates about things which cannot be otherwise than they are […] and the man to whom we should give the name of Good in Counsel, simply and without modification, is he who in the way of calculation has a capacity for attaining that of practical goods which is the best for Man.
>
> (Aristotle 1910: 173)

Thus, Aristotle's famous "All men by nature desire to know", became identified with the pursuit of disinterested knowledge. The philosopher pursues knowledge for its own sake and not for any practical purpose, so the division between theoretical philosophy and practical philosophy was established. His division between intellectual virtues and ethical virtues and the corresponding ways of life was a way to do so, though for him the contemplative life was superior to the political life in terms of happiness. In any case, the separation between the sphere of knowledge and the sphere of ethics would be assumed from then onwards. In the case of Plato, we do not have enough evidence to claim so, but he did establish an exclusive relationship between truth and knowledge, which may have contributed to it. Indeed, *the love of truth was understood as the love of wisdom* when it could have been understood in a broader sense as the *love of knowing*. But, instead, wisdom as a virtue became identified with "the part with which we learn", "which is forever striving to know where the truth lies, and cares for money and reputation less than the others", that is, the part of the soul which desires to know and which was later distinguished by Aristotle from practical wisdom. That division within the soul seems to have been responsible for separating ethics from knowledge and for making moral philosophy a branch of practical philosophy related to politics. The problem was that separating the "desire to know" from ethics meant that the love of wisdom was going to be separated from practical affairs as a disinterested desire to pursue truth for its own sake.

Of course, Plato's identification of the love of truth with the love of wisdom was well-intentioned because, like Socrates, he was very critical of the teachers of wisdom, the sophists, who were in the business of earning a living and building a reputation for themselves from teaching the art of persuasion to those who could pay their fees. The sophists charged money for teaching how to win an argument and convince their audience, a very useful skill at a time when Athens was consolidating its pre-eminence among the Greek city-states and needed trained politicians and citizens. In the midst of this, appeared Socrates who was opposed to the *modus vivendi* of the sophists and was willing to educate a new generation of people who would serve the city wisely. Rather than selling wisdom to make money, he tried to instil in them the love of truth, and, though he was less successful than the sophists in carrying out his mission, it paid off with Plato. Indeed, his most gifted student would become not a politician but a true philosopher who opened a school, the *Academy* (c 387 BC–84 BC), that would not charge its students for teaching wisdom and which would later paid off with Aristotle, who attended it for 20 years and created his own school, the *Lyceum* (334 BC–86 BC). This sequence of schools is just to point out that the *love of truth* moved true educators to create human systems that would realize the value of knowledge through learning and teaching knowledge to others. It was very difficult to realize that value without institutions whose purpose constributed to realize the value of knowing in the human world. Fortunately, it is said about Plato that after having failed to convince the tyrant of Syracuse to apply his ideas and ending up being sold as a slave, he decided to invest the money that someone paid to recover his freedom in building maybe the first school of higher learning. Because Anniceris, the Cyrenaic philosopher who bought his freedom back, did not want the money back. We do not know if the story is true, but the foundation of that human system contributed to the realization of the value of knowledge in the human world which continues to this day with schools and universities worldwide.

In his *Memoralia* (1899), Xenophon describes Socrates as someone that loved educating and could not bear the sophists who traded their wisdom for money, going as far as comparing it with prostitution. Instead, he wanted his followers to be educated in those things that they were most suited to and "he taught them himself, whatever he knew", "and what he had not sufficient knowledge to teach, he took them to those who knew, to learn" (Xenophon 1899: 183). He understood that his true vocation was to devote his life to educating others because education was "the greatest good for human beings". In the *Apology of Socrates* (1892), Plato also says that his occupation was not "corrupting" the youth, one of the allegations made against him in the trial that would condemn him to death, but making them better through "wisdom and truth". "For I do nothing but go about persuading you all, old and young alike, not to care for your persons or your property, but first and chiefly to care about the greatest improvement of the soul" (Plato 1892: 30). Plato was very critical of the sophists in the *Protagoras* because they sold goods without knowing which teachings were "helpful or harmful to the soul". It is dangerous to buy something that could corrupt the soul. As a matter of fact, in that same dialogue in which he questions what sophists claim to be doing for others, Socrates seems to think that you cannot teach virtue. Indeed, the love of wisdom has to be distinguished from the acquisition of knowledge. You can learn lots of knowledge without being moved by the love of truth. Instead, sophists were moved by its countervalue, *scepticism*, because they had contempt for truth inasmuch as they used whatever "knowledge" they could find to prove or disprove something without caring about wisdom. The claim that "Man is the measure of all things", attributed to Protagoras, shows their contempt for truth because everything can be true or false, it depends on the ability of the sophist to persuade others which reinforces *ignorance* in our world.

We have reached upon the countervalue of the *value of knowing* through one of its forms, scepticism, which is reinforced by human beings who have contempt for truth and, therefore, are do not care about truth. However, the only way to counteract countervalues is by realizing the values to which they are related. In the case of Socrates, it meant fighting the spread of scepticism with regards to human matters that still prevails in our time given our state of ignorance regarding our understanding of the human world. We know more about the universe than about ourselves and our world. According to Xenophon (1899), Socrates exhorted his students to know themselves, by which he meant that they had to discover what they were good at in order to contribute to the city, "it was his care that his followers should be competently qualified for employments suited to their powers" (Xenophon 1899: 183). But the question concerning the human being itself was beyond his reach because to provide that concept requires a world-hypothesis, something that had to wait for Plato. Of course, at that time the sophists took advantage of the prevailing ignorance with regards to the cosmos and extended that scepticism to human matters, which, of course, aroused the philosopher's contempt for the sophists because the genuine ones are fond of truth and wisdom and cannot accept that form of ignorance. Even though they did not contribute much to our knowledge of human matters, at least they did their best to fight the deniers of true knowledge. "The sophists speak to deceive and write for their own profit, and they never benefit anyone in any way", we are told by Xenophon in the *Cynegeticus*.

In fact, the sellers of wisdom took scepticism to extremes because, besides not caring about truth, they also denied the possibility of knowledge itself. Gorgias, another sophist to whom Plato devoted another dialogue, exemplified this radical position. He claimed, according to Sextus Empiricus (1935), "that nothing exists; second, that even if something existed, it is inapprehensible to man; and third, that even if it was apprehensible, it cannot be communicated nor explained to others" (Empiricus 1935: 41). Indeed, if we cannot discover and trasmit knowledge, then we cannot learn or teach true knowledge. On the contrary, it has been the love of knowing that has moved philosophers and scientists to pursue knowledge because they care about others. We owe the realization of the value of knowledge to them and to the true educators who can teach that knowledge to others.

The question we should ask ourselves is whether the love of truth and wisdom is necessarily correlated to the discovery of true knowledge. In the case of the educators, who are not themselves philosophers or scientists, it is clear that they may not discover knowledge but they have the same desire to know. Indeed, education systems can realize the value of knowing without necessarily discovering any knowledge, as long as they are able to teach knowledge to others. In fact, it is even possible for them to teach something that is not knowledge with a passion that will awaken the desire to know in others. The history of philosophy is a good example of such a quest for knowledge, because most of what true philosophers have tried to teach us is not knowledge as such, which does not mean that there is no knowledge in philosophy. Philosophers can discover true knowledge, like scientists, but not in the same way because philosophy and science are different domains of knowledge. The domain of science, as I have tried to demonstrate with my work (Pretel-Wilson 2020), is grounded on the domain of philosophy whose advancement makes the progress of science possible. However, even though the desire to know is not necessarily correlated with the discovery of knowledge, the advancement of knowledge depends on the love of truth, just like the adquisition of knowledge depends on the desire to know. So what drives the lovers of truth is the *desire to know that contributes to the realization of the value of knowledge through the endless pursuit of knowledge.*

Someone could wonder if there is any need of learning knowledge if the pursuit of knowledge is enough to realize the value of knowledge, and I would answer in the negative because pursuing knowledge includes both discovering *and* teaching knowledge. In other words, if education systems did not teach any knowledge, then they could never realize the *value of knowledge* because without knowledge there is no true learning. I have sometimes wondered about the difference between teaching and researching when it comes to human systems given that schools are focused on teaching and universities combine teaching and researching. Indeed, it is tempting to make a split between colleges and research universities, but both human systems are education systems that pursue knowledge because to discover knowledge is to learn knowledge that can be taught to others. So we have the first essential component of the first fundamental law of ethics.

Now I can introduce the other essential components to show that the love of truth cannot be reduced to the love of wisdom given that it is also present in theother values that belong to the same group. Having said that, the philosopher who laid stress on our next value, could not avoid relating it exclusively to knowledge by making it a way to approach truth. Popper termed *criticism* that value, though *fallibility* seems more appropriate, and its countervalue, *dogmatism*. However, he understood them in terms of the logic of discovery of scientific knowledge suggesting a correspondence with conjectures (dogmatism) and refutations (criticism). The *value of fallibility* is another form of knowing related to the pursuit of knowledge but not exclusively, and dogmatism is another form of ignorance besides scepticism. Let me start by comparing the two countervalues I have introduced so far. Scepticism, as we know, moves people who do not care about truth and deny possibility of knowledge. Of course, nowadays we have an updated form of scepticism because it has already been refuted with the discovery of scientific knowledge. In any case, even if it cannot claim an absolute doubt in all the domains of knowledge, it still uses an unfounded scepticism to generate doubts in some matters where there is a general consensus among scientists and, by doing so, reinforces ignorance in the human world. In contrast, dogmatism moves people who have contempt for all knowledge that contradicts what they believe to be absolutely true. Scepticism, on the other hand, does not mind questioning everything because it does not hold anything dear.

In Aristotelian terms, we could say that dogmatism seems to be the opposite of scepticism in a continuum between absolute truth and absolute doubt and criticism or fallibility the middle ground. Of course, Popper ([1963] 1972) rejected scepticism as "the prejudice of seeking to prove everything" and endorsed dogmatism because he saw an evolutionary relationship between the "dogmatic attitude" and the "critical attitude", between the "pseudo-scientific" or "pre-scientific attitude" and the "scientific attitude". However, "this critical attitude is not so much opposed to the dogmatic attitude as super-imposed upon it: criticism must be directed against [...] dogmatic beliefs" (Popper [1963] 1972: 50). Indeed, the first phase of the logic of discovery is about holding those dogmatic beliefs as "conjectures" and the second phase is about "refutations", that is, questioning them and discarding them. "The formation of a theory or conjecture has always a 'dogmatic', and often a 'critical', phase" (Popper 1992: 45). In any case, that dogmatic attitude is defined in opposition to the critical attitude and, as we know, a countervalue is not the antithesis of a value, so criticism might not be the value we are looking for but something whose countervalue is dogmatism. Furthermore, there is no necessary relationship between criticism and learning as the sophists have taught us, whereas fallibility suggests "that *we can learn from our mistakes*" (emphasis in the original), which, according to Popper's preface to the first edition, is the thesis that informs *Conjectures and Refutation* ([1963] 1972).

Indeed, what moves people that love to learn is not criticism *per se* but being better at what they do because *human systems realize the value of fallibility through doing something for others*. It is tempting to think that all learning comes from doing something wrong, that "*all* our knowledge grows *only* through the correcting of our mistakes", as Popper wrote in preface to the second edition (1965), that is, "the method of trial and error". But do we learn from getting things wrong or from getting things right? Yes, from both. Fallibility implies that we can learn from discarding the things that do not work *and* from improving the things that could be done better. This is where we start to see that the love of truth is also related to pursuing something other than knowledge. Fallibility is realized by human systems through *doing something right for others* on a regular basis. This is important because one thing is to do something well and the other to do something better. This implies that they learn from asking and listening to those to whom they serve and from seeing and reflecting on what could be done better. Dogmatism, on the other hand, is not open to criticism and does not learn through what it does for others. On the contrary, humans moved by dogmatism are immune to criticism.

The value of knowledge is related to the value of fallibility because, as Popper argued, we discover knowledge by questioning scientific hypotheses. We have to assume that the hypothesis is correct and derive the observable consequences or predictions to see if it is the case or not? We must listen to what the world has to say about it. If the world does not refute the hypothesis, then we must take it provisionally because we may find a better hypothesis to explain the facts or even discard it if the observations disprove the hypothesis. If it cannot be refuted, it will be our best available explanation at the time but not the definitive answer, that is, the absolute truth, because it can be questioned in the future and replaced by a better explanation. This is not something exclusive of science because philosophy also has to question its own hypotheses even if there are no better answers to its fundamental questions because they are never definitive answers. The pursuit of knowledge depends on always being able to provide better explanations and, hence, rejecting dogmatism on the grounds that there are no absolute truths that cannot be questioned. However, questioning a hypothesis is not the same as criticism for its own sake since the lovers of truth do not reject a hypothesis for the sake of rejecting it but for the sake of finding a better explanation to replace it. Questioning for the sake of questioning without having anything better to hold on to, at least provisionally, is scepticism. This is why Popper (1992) did not want to get rid of dogmatism completely and thought that it was necessary for the development of theories as "some degree of dogmatism is fruitful". "Most of the great theories which are among the supreme achievements of the human mind are the offsprings of earlier dogmas, plus criticism" (Popper 1992: 48).

Of course, I would not say that holding on to something for the lack of having something better to replace it with is dogmatism but something else, because people who are moved by the love of truth do not stick to something unless there is no alternative and will not mind replacing it when they find something better.

This relationship between the value of knowledge and the value of fallibility is also the case between the *value of veracity* and the *value of honesty*, which are the other essential components we are missing to complete the first fundamental law of ethics and which are also realized by human systems who also love truth. Our next two values are so closely related to each other that they tend to be confused. In fact, Aristotle (1910) was the first philosopher to do so when relating veracity or truthfulness to two different countervalues that are related to two different values. He understood veracity as a virtue between two vices, dissimulation and boastfulness, depending on whether there is reservation or exaggeration, "the two who depart from strict truth are both blameable, but especially the Exaggerator" (Aristotle 1910: 118). The exaggerator or deceiver lies whereas the dissimulator or hypocritical hides the truth and, in so doing, they reinforce the countervalues of *lying* and *hypocrisy* in the human world. Aristotle did not notice the difference between them because he saw the two extremes as different forms of falsehood, falling short of the truth and beyond the truth, because he was looking for contrary vices that were opposed to the virtue of truthfulness, namely, telling the truth. But as we know, that is not the way to define values and countervalues because, to repeat it once again, they are not defined in opposition to each other, like true and false, so there is no point in "speak[ing] of those who are True or False, alike in their words and deeds and in the claims which they advance" (Aristotle 1910). Instead, people that do not lie are moved by the love of truth in the same way as people who do not hide the truth also love the truth, but they realize two different values by doing so while pursuing a purpose. Likewise, those people that do otherwise have contempt for truth and reinforce ignorance in the human world through *what they tell and do not tell others*. To be fair with Aristotle, his "Reserved" is not our dissimulator but someone who "pretends to less" than they are entitled to, someone "who depreciates their own qualities", as opposed to someone who "pretends to more", the "Boaster". To avoid confusing the reserved with modesty and its opposite with vanity, it is better to understand hypocrisy as pretending, something very human that we do not find in the animal kigndom, according to Schopenhauer:

> [A]nimals are not capable either of purpose or of dissimulation; they have nothing in reserve. In this respect, the dog is related to the man as a glass tumbler is to a metal one and this greatly helps to endear the dog so much to us. It affords us great pleasure to see simply and openly displayed in him all those inclinations and emotions that in ourselves we so often conceal.
>
> (Schopenhauer [1818] 1968: 61)

Indeed, animals are transparent like a glass of water as they have nothing to hide but not because they are honest, but because they react to stimuli from their environment and cannot hide their moods. In contrast, humans can hide their emotions and their true purposes behind what they say and do and, in doing so, reinforce hypocrisy in the human world. What hypocrisy and lying have in common is that they *lie to someone, either to themselves or to others,* and that is why they are different forms of ignorance. To my knowledge, Schopenhauer ([1851] 1900, 1915) was the first philosopher to notice the difference between the values to which those countervalues are related. Honesty is discussed in relation to the "post-Kantian school" and veracity in relation to a dialogue about religion. According to him, "Fichte, Schelling and Hegel are not philosophers". "They are mere sophists; they wanted to seem and not to be, and have sought, not the truth, but their own interest and advancement in the world" (Schopenhauer [1851] 1900: 22). To be a philosopher "one should love the truth earnestly and with one's whole heart, and therefore unconditionally, without reserve, above all else, and in case of necessity, despite everything" (Schopenhauer [1851] 1900: 23). However, according to Schopenhauer, that was only possible in philosophy, "it will not get us far in the world". For him, honesty was only possible in philosophy because you cannot serve the world and truth at the same time and any attempt to do so will lead to hypocrisy. Again, the love of truth cannot be related to wisdom exclusively and separated from the world as something that does not care about truth because humans realize honesty in the human world through what they do for others. Anything that people do can be done with honesty even if the purpose of what they do is not to pursue knowledge. A great example of intellectual honesty was provided by Fregue, the father of modern logic, when he acknowledged a mistake discovered by Bertrand Russell that proved that it was impossible to reduce mathematics to logic and just when he was correcting the proofs of *The Basic Laws of Arithmetic* (1903).

> Hardly anything more unfortunate can befall a scientific writer than to have one of the foundations of his edifice shaken after the work is finished. This was the position I was placed in by a letter of Mr Bertrand Russell, just when the printing of this volume was nearing its completion.
>
> (Fregue 1903: Appendix)

The other unconditional love of truth that Schopenhauer mentions in the context of a dialogue between Demopheles and Philatetes, who represent Religion and Truth, is veracity. "I respect truth everywhere, and so I can't respect what is opposed to it" (Schopenhauer [1851] 1915: 3). Demopheles is trying to convince Philatetes that the message of religion can only be understood by those who are initiated and, therefore, it has to use allegories suitable to those who are cannot, because "a parable or myth, is all that would be of any service to them" (Schopenhauer [1851] 1915: 12). However, for Philatetes, "the naked truth must be so simple and intelligible that it can be imparted to all in its true form, without any ad mixture of myth and fable, without disguising it in the form of religion" (Schopenhauer [1851] 1915: 21). Demopheles tries to argue that the lies that Philatetes has contempt for have "practical aims", which "are in every respect superior to theoretical [purposes]", and attacking them could "destroy its invaluable influence on the hearts and morals of mankind".

> Instead of insisting on that with pedantic obstinacy, look at its great achievements in the practical sphere, its furtherance of good and kindly feelings, its guidance in conduct, the support and consolation it gives to suffering humanity in life and death.
>
> (Schopenhauer [1851] 1915: 28)

For Philatetes,

> The truth, my dear sir, is the only safe thing; the truth alone remains steadfast and trusty; it is the only solid consolation; it is the indestructible diamond. [...] Error always does harm; sooner or later it will bring mischief to the man who harbors it. Then give up

> deceiving people; confess ignorance of what you don't know, and leave everyone to form his own articles of faith for himself.
>
> (Schopenhauer [1851] 1915: 29)

Little did Schopenhauer know that that contempt for lying was grounded on ethics and comes from the love of truth that moves human systems to realize the value of veracity. We have come full circle in our understanding of the first fundamental law of ethics that moves all peoples and civilizations in the universe. Indeed, humans realize the *value of knowing* through what they pursue for others on a regular basis moved by the love of truth and contempt for different forms of ignorance that are reinforced in the human world through what humans pursue. That understanding should also help us get rid of Aristotle's artificial division between the theoretical and the practical, which Schopenhauer was alluding to, because ethics is a theoretical-empirical science. Scheler and Hartmann also made an attempt to hold on to that classical division in philosophy. The last moral philosophers claimed that "truth does not belong among the values", according to Scheler, that it is a "theoretical value", like "evidence" and "knowledge", according to Hartmann, who distinguished it from veracity, the correspondence between words and convictions, which was a "moral value". They did not realize that human systems realize the value of knowing through doing other things besides pursuing knowledge, moved by the love of truth. And yet, Scheler (1913–1916) argued that though "Truth is *not* a value", "it makes good sense to attribute a value to acts of *searching*, the investigation into truth" and included the "*pure cognition of truth*, whose realization is sought in philosophy" (Scheler [1913–1916] 1973: 108), among the "spiritual values". It did not occur to them that *the realizing the value of knowing through what people do to counteract ignorance was the first law of ethics.*

Let me finish this chapter answering the question as to who are the people that realize the values of veracity and honesty in the human world. Well, it seems that they are the same who have *contempt for lying* because the people that love the truth hold it dear. Those values are realized by those human systems whose purpose is to *provide accurate information to others on a regular basis*. The media would be a good candidate if it was not that sometimes it is prone to bias and propaganda, but good journalism is not. We also have trust in companies because we assume they comply with the rules and regulations.

Unfortunately, companies are also generating information on a regular basis through paid publicity or *advertising* whose only purpose is to persuade more people to buy their products even if they do not need them and, in so doing, they reinforce ignorance in the human world because people should not be mislead in buying what they do not need. Companies should provide accurate information to consumers so they can decide if they need that product to satisfy their needs or not. Indeed, the value of a product depends on the our knowledge that it contributes to satisfy our needs and, therefore, companies should provide that right information so people could make informed decisions as to what they need to buy. If companies provide accurate information to others, instead of using misleading information to attract buyers, they will be realizing the value of veracity and honesty every day. And the same goes for non-profit organizations that want to receive money from donors by providing accurate information about what they do for others and how their money contributes to the purpose they serve without trying to make them feel good about funding

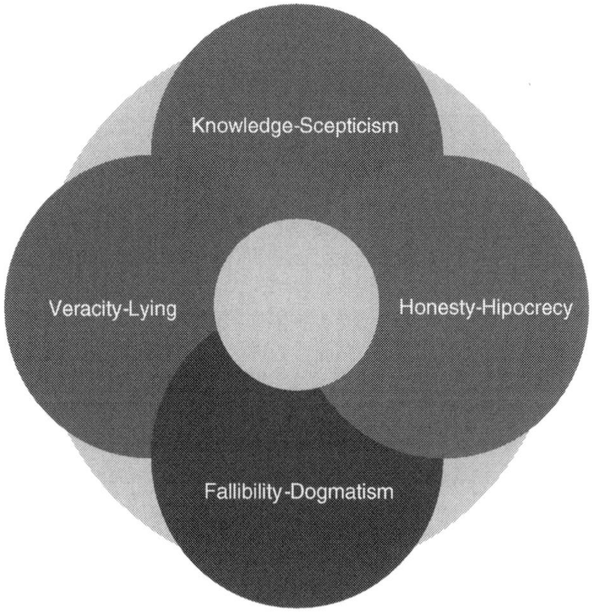

Figure 6.1 The First Fundamental Law of Ethics: Knowing-Ignorance.

their cause and bad if they do not. Both companies and non-profit organizations can educate others to make better choices.

We can conclude this chapter with the set of values and countervalues that constitute the first fundamental law of ethics (Figure 6.1).

References

Aristotle (1910). *Nicomachean Ethics*, trans. G. H. Lewes. London: George Routledge.

Frege, G. (1903). *Grundgesetze der Arithmetik,* Band II. Jena: Hermann Pohle.

Plato (1892). *Apology of Socrates*, trans. B. Jowett. London: Humphrey Milford: Oxford University Press.

Popper, K. (1972). *Conjectures and Refutations*: *The Growth of Scientific Knowledge.* London: Routledge and Kegan Paul.

Popper, K. (1992). *Unended Quest: An Intellectual Autobiography*. London: Routledge.

Pretel-Wilson, M. (2020). *Utopics: The Unification of Human Science*. Cham: Springer.

Scheler, M. (1973). *Formalism in Ethics and Non-Formal Ethics of Values*: *A New Attempt toward the Foundation of an Ethical Personalism.* Evanston: Northwestern University Press.

Schopenhauer, A. (1968). *The World as Will and Representation*. New York: Dover Publications.

Schopenhauer, A. (1900). *Selected Essays of Arthur Schopenhauer*. London: George Bell.

Schopenhauer, A. (1915). *Religion*: *A Dialogue and Other Essay*. London: George Allen & Unwin.

Sextus Empiricus (1935). *Against the Logicians*, trans. R. G. Bury. London: William Heinemann.

Xenophon (1899). *Memoralia*, trans. J. S. Watson. Philadelphia, NJ: David McKay.

7 The Value of Life

The next law of ethics could not be appreciated by moral philosophers because it was overshadowed by a dualism I have already discussed that made them underestimate life. Plato (1970) had started that trend when conceiving the soul as divided into three parts, in which the appetitive part was prone to the *pleasures of the body* and, hence, had to be controlled by the virtue of *temperance* or moderation, in the same way as the irascible part needed to show *courage* and the rational part *wisdom*. That meant that the soul was trapped into a body, which is how Plato conceived the human being, and prone to weaknesses because of its exposure to bodily pleasures. Of course, he could not deny them because he understood that a healthy life depended on satisfying the needs of the body, just like a healthy city depended on satisfying the needs of its citizens. However, he also thought that without controlling the appetitive part, the city could become an insatiable "luxurious city" that would lead to wars with its neighbours. So if it wanted to become a just city, its citizens had to live with moderation, avoiding living a life of excess. Furthermore, a healthy life also depended on nurturing a healthy body through gymnastics, which was part of his system of education for "the guardians of the city" who had to protect it from possible invaders and from intemperate citizens that could cause a rebellion. To avoid this outcome, the irascible part had to be aligned with the rational part, who ruled the city with wisdom: "is it not befitting that the rational part should rule, inasmuch as it is wise and has foresight for the whole soul, and that the spirited part should be its subject and ally?" (Plato 1970: 131). Following Plato's narrative of the coming into being of the perfect city, it was natural to associate the root cause of injustice with the appetitive part

DOI: 10.4324/9781003649496-9

of the soul responsible for "those passions which are most responsible for all the evils that come upon cities and the men that dwell in them" (Plato 1970: 53).

Consequently, the body became associated with vice and, although a happy life depended on a healthy body, the goods of the soul were ranked higher than the goods of the body. According to Aristotle (1910), the contemplative life of the philosopher and the prudent life of a politician were superior to a "life of sensual enjoyment". However, he insisted that some pleasures and external goods, like fortune and glory, were necessary for a happy life and, in order not to dismiss them, he claimed that "wisdom, virtue and pleasure" are the pillars of happiness. Instead of three parts, the soul now had two parts, one that ruled and another that obeyed and, accordingly, there were two types of virtues, the intellectual and the moral. Unfortunately, health became a good that was necessary for happiness, like fortune, glory, and friendship, but not a virtue. But he kept temperance because it was a virtue between two vices related to pleasure, intemperance as the excess and insensibility as the defect. Temperance "consists in resisting, through reason, the passions and desires that one feels in its soul" and intemperance "gives in to pleasure, and succumbs with a contemptible sort of weakness". In other words, intemperance was being seduced by the pleasures of the body, which he identified with the senses of touch and taste, most probably because he saw sight and hearing as connected to the pleasures of the soul, as we know from his *Metaphysics*. In any case, the body became a burden to ethics to the point that Kant decided to separate happiness from ethics and to argue that *life has no value* unless we live "with dignity, that is, without violating the dignity of humanity" in our person. Furthermore, the person is the only being in the universe that has an absolute value and all other *things are mere means*.

The underestimation of life was the consequence of body-soul dualism in which the appetitive part became a concern for ethics. This created a strong reaction from Nietzsche ([1886] 1917), who argued that ethics could not be separated from life because the appetitive part was intrinsic to ethics, which had nothing to do with reason but with power. What is most useful to a "will to power" is not truth but something that fulfils its "physiological demands", an "opinion is life-furthering, live-preserving, species-preserving", which is more important than truth because false opinions "are the most indispensable for us". Thus, we have "To recognize untruth as the condition of life" (Nietzsche [1886] 1917: 4). But wait a moment, is "untruth" what preserves life? Hasn't it been the discoveries in medicine what has contributed the most to the enhancement of life? Furthermore, is all life instinctive to the core, or are instincts what is passed on by our genetic code and has nothing to do with ethics? Rather than questioning the details of what he said, we have to understand that Nietzsche was reacting against the concept of a rational human being. And the future generation of philosophers, to whom he was addressing *Beyond Good and Evil* (1886), understood at last was had been missing. The *value of life* was going to enter the field of moral philosophy with full legitimacy. Indeed, Scheler and Hartmann included vital values in their hierarchy of values distinguishing them as moral values from "the agreeable" and "the useful".

In the case of Scheler (1913–1916), he believed that each value-modality was correlated to a type of feeling and what "vital feelings" experienced in "living things" where "the values of noble and vulgar" or vital values that "belong to the sphere denoted by "weal" or "well-being". As a phenomenologist, he gives a list of "feelings of life", such as "the feeling of health and illness", and of emotional reactions, such as "being glad about" or "being sad about". "Life is an object of valuation not only as this or that living being (insofar as beings incorporate more or less life) but also as an *essence*" (Scheler [1913–1916] 1973: 275–276). According to Scheler, this life is not the same life that the scientist perceives under the "natural view of the world", that is, the "naturalistic attitude" that Husserl wanted the phenomenologist to escape from, because it "could never *know* life *as an object*", "could not *value* life itself". In contrast, the phenomenological attitude can experience the value of life whose access depends on acts "feeling, loving, hating", but whose being is independent of the bearer of values, which is what he meant when he said that what those emotional acts have access to an essence that is independent from our experience. Despite having or claiming to have discovered the value of life through emotional acts, he could not accept Nietzsche's "*presupposition* that life is the highest value" (Scheler [1913–1916] 1973: 285). For Hartmann ([1926] 2011), the value of life was so basic that it provided the ground of moral values in as much as the living being is the ontological foundation of the psychic being and, indirectly, of "the moral being and the carrier of the moral value", the person. Indeed, he distinguished between different strata of reality that rested on top of each other, so the person was grounded on the biological stratum. Like Nietzsche, he seems to understand life as "natural and instinctive", "The value of what is animal in man". But neither Scheler nor Hartmann understood the difference between living beings and sentient beings, as they could not see beyond the body-soul dualism.

Given the relationship between values and countervalues, *life* is that which is loved by someone who has contempt for *suffering*. But how do we define the love of life? It is easy to say that we value our own life inasmuch as we love our own happiness, but this does not mean that we love other lives and want other people to be happy. People that care about their own wellbeing, might not be interested in the wellbeing of others. But it is rare to find people who do not care about anybody but themselves. Most if not all, people care about *their loved ones* and are ready to do whatever it takes to reduce or prevent their suffering and, equally, to make them happy. The love of life can also show in people that want to look after themselves in order to remain healthy as long as possible and to be around their loved ones. But saying goodbye to our loved ones is, in fact, the love of life in its most traumatic form as it is the highest form of suffering, losing someone we love. So nobody is immune to the love of life no matter how insensitive they are to others, in the same way as nobody can be immune to the suffering of others. We feel pity for people in need, for those who have lost their loved ones, for those who are suffering or have not had an easy life, because we cannot bear their suffering no matter if they are our loved ones or not. However, this is no longer the love of life but compassion, an emotion correlated with suffering. *The love of life is the love of the opportunity to live,* which suffering or death takes away. People who love life cannot bear the loss of that gift and do whatever it takes to honour that priceless opportunity. They treat their *life as a miracle, as something that will never repeat again,* as an opportunity that does not last forever and that could be gone at any time. That opportunity that has been "given" to us is what moves the lovers of life in their quest to find something to do with it. But that opportunity to live is also a miracle granted to others, so the love of life is not exclusive to our own life nor to the life of our loved ones. Therefore, the love of life is the *love of the opportunity to live of every living being that has been granted that miracle.*

As I said earlier, the value of life is related to the countervalue of suffering but also to *death* because someone who loves life has contempt for suffering *and* death. If death was absent from the countervalue, the relationship to the value of life would be incomplete because the opportunity to live is lost the moment that death comes knocking on the door. Indeed, death is a concern for anyone who loves life and knows that that miracle is a countdown whose time cannot be stopped or restarted again. That contempt for death always accompanies the lover of life, no matter how healthy or happy they are or how often they think about it, because they are constantly remained by those around them who have had to say goodbye like themselves one day. If this is the case, we have to disagree with Spinoza's (1677) proposition: "A free man thinks of nothing less than of death" and hence he "is not led by the fear of death" (Spinoza [1677] 1970: 187). People who are moved by the love of life are indeed concerned about death, maybe not every day but it bothers them to know that it will happen one day and they fear that day. This fear of death is something very human and not the same as the emotion of fear that animals experience when they face a danger that can harm them. Animals do not fear death, even if they are exposed to life-threatening situations and experience the pain of losing a companion or an offspring and even someone that was part of their lives. Instead, what they fear when they encounter something potentially harmful is the pain that might result from that threat which they anticipate. Humans also have the fear of an impending threat that can kill them, but what they fear when they fear death is dying one day. The fear of death results from the love of life of people who are not ready or prepared to say goodbye because they have many things they want to do before that day eventually comes.

As we already know, the realization of a value always implies pursuing a purpose that realizes that value, and pursuing a purpose is doing something for someone. In the case of the value of life, we do not realize that value by taking care of our own life but taking care of that miracle granted to living beings like us. Of course, we can only take care of life if we love that miracle. Some people can even love that miracle by making it their life purpose to take care of others on a regular basis. In the history of moral philosophy, that came second to taking care of ourselves, the duty to oneself before the duty towards others. Caring about ourselves meant the pursuit of our own happiness to which the "contemplative life" contributed more than devoting our life to other things, according to Aristotle, such as helping others, for instance. But isn't the love of life as unconditional as the love of truth? The lover of life is ready to save the lives of those who love life *but also* of those who have contempt for life. For Spinoza, taking care of our bodies was important because virtue resided in preserving our being. "No one endeavours to preserve his being for the sake of anything else". Caring about our being for the sake of ourselves was the most ethical thing to do. "Therefore to act according to virtue is nothing else in us than to act, to live, and preserve our being according to the guidance of reason, on the basis of seeking what is useful to oneself" (Spinoza [1677] 1970: 158).

What about the other living beings that have also been granted that miracle? Of course, for Aristotle, animals were the quintessence of what we had to keep out of ethics as they are not moved by reason. He contrasted the philosophical life and the political life to the "life of pleasure and enjoyment", "the life of the beasts". But he could equally have said the "life of pain and suffering" because "the fact of living, so far as animals are concerned, is characterized generally by the power of feeling". He did not say much about how we had to treat animals, besides saying that we cannot live like them. However, Spinoza was quite clear about it, "whatever exists in nature besides man", "we should preserve or destroy it according to our various needs".

> I do not deny that beasts feel; but I deny that on that account we should not consult our necessity and use them as much as we wish and treat them as we will, since they do not agree with us in nature, and their emotions are in nature different from human emotions.
>
> (Spinoza [1677] 1970: 167)

But what if we discovered that humans share the same emotions? Well, to do so, we have to get rid of the body-soul dualism and replace it with the anatomy of the universe I have introduced in chapter 4. The fundamental science of ethics has to acknowledge that human beings and human systems cannot cancel out the spatial-temporal dualities that constitute cybernetic (sentient) systems, in the same way as cybernetic systems presuppose the spatio-temporal dualities that constitute biological and physical systems. Indeed, *our constitution presupposes the cybernetic constitution and, thus, we share the same emotions experienced by animals.* So Aristotle was not very convincing when he assumed that pleasure and pain are felt by the soul, because he could equally have said that they are felt by the body. Furthermore, he distinguished between the pleasures of the body and the pleasure of the soul, linking the former to the senses of touch and taste and the latter to sight and hearing, not knowing where to fit smell in that dualism. That division was done intentionally to separate eating and drinking, and "the so-called pleasures of love", from intellectual activities, such as "contemplation [which] constitutes what is most agreeable and excellent", according to his *Metaphysics*. Knowledge is pursued for its own sake whereas the pleasures of the body are pursued to satisfy a need: but of the body or of the soul? The solution to that limited understanding of the human being is to get rid of the body-soul prejudice once and for all. Pleasure and pain, in addition to d all the other emotions that human beings share with animals, and that we will discuss in the last chapter of the book, come from the cybernetic (sentient) constitution which our constitution presupposes. But, most importantly, all suffering presupposes that constitution and, therefore, *the contempt for animals is also the contempt for life because they are also living beings.*

We cannot blame all moral philosophers for being insensitive towards animals. Hume (1739), for instance, saw that we share much in common with them. "The three relations of resemblance, contiguity and causation operate in the same manner upon beasts as upon human creatures" (Hume [1739] 2000: 213), that is, we share the same "principles" that explain how the mind works. In particular, "the relation of impressions, sufficient to convince us, that there is a union of certain affections with each other in the inferior species of creatures as well as in the superior" (Hume [1739] 2000), that is, we share the same emotions, such as joy and kindness, grief and anger, in addition to pleasure and pain. However, Hume went a bit too far in his claim that the "*indirect* passions", which "arise from a double relation of impressions and ideas", such as "pride or humility, love or hatred", also apply to "the whole animal creation". In the same vein, Smith believed that animals share the "passions of gratitude and resentment" which are excited by pleasure and pain. As we saw earlier, similarly to Spinoza, Kant considered animals as things that could be treated as mere means for our purposes. In the *Conjectural Beginning of Human History* (1786) he even made use of the *Genesis* to point that the skin of the ship was given to us by nature "for my use, not for yours", so animals had to be treated by a human being "as mere means and tools to whatever end he pleased". However, Kant was a bit more kind to animals in his *Lectures on Ethics* (1788–1789) in which he recognized an indirect duty towards them because they have an analogous nature to ours and we must treat them with kindness, "for he who behaves cruelly towards them also has a hardened heart towards his fellow men". And he was against performing experiments on animals unless we could benefit from them, "but it cannot be tolerated in any way that they are practiced as a game" (Kant [1788–1789] 1988: 288). Finally, Schopenhauer was very critical of Spinoza for his contempt for animals but excused him blaming it on Descartes' split of "nature into spirit and matter, thinking and extended substance".

Schopenhauer argued that we cannot mistreat animals on the ground that we share the same essence that manifests itself in different forms in the world of phenomena, but is it, deep down, one and the same thing, "the thing in itself" or Will. The pain that we inflict on animals can only be justified by the pain that humans avoid from eating their meat or using their force, "the will to live as a whole thus endures less suffering than if the reverse occurred". However, we would do injustice to them if we overexploited them beyond what is necessary. Like Kant, Schopenhauer ([1844] 1909) disliked the experimentation on animals, especially on higher animals, given that the degrees of suffering increases and reaches its peak in humans. "[T]he brutes have infinitely less to *suffer* than we have, because they know no other pains but those which the *present* directly brings" (Schopenhauer [1844] 1909: 230). Their suffering is not like ours because it does not include things of the past nor of the future to which we have to add "that which is merely possible". The more perfect is the manifestation of the Will, the higher is the corresponding level of suffering; from plants that do not feel any pain to vertebrates with a nervous system that can experience it, and it reaches its zenith in the case of humans with a developed intelect that suffers much moregence. But all life shares the same essence, "the *eternal essence that, like in us, also lives in all animals*", "*the animal* is in essence and in the main exactly the same as us". Therefore, we should be more considerate towards them instead of mistreating for not having a rational and immortal soul, as Descartes thought. Schopenhauer ([1851] 2009) blames Christianity for having inherited from Judaism "the view that regards the animal as something manufactured for man's use". The belief that God created the earth and gave us all the animals as things over which we had a right and could do what we pleased with them. And he even granted us the right to name them as a sign of our mastery over animals. In contrast, Buddhism acknowledges the affinity of humans with nature, "but first and foremost with animal nature". According to him, the East is morally superior to the West, "since Brahmaninsm and Buddhism do not limit their precepts to 'one's neighbor' but take under their protection 'all living beings'" (Schopenhauer [1851] 2009: 382, 388).

The people who are moved by the love of life care about the well-being of others not just their loved ones or other human beings, like the Good Samaritan. The love of life cares about all life, not just human life, and about the wellbeing of the Earth as a living ecosystem. However, that living being is constituted by an entangled web of species that cannot live without each other and that holds together in a stable equilibrium that one single species is disrupting. That ecosystem is the miracle of life on Earth that has evolved for millions of years adapting to all the changes in its environment and spreading throughout the different habitats that host the tree of life. Well, as I said, one single species is now upsetting that stable equilibrium not only by changing the conditions of the physical systems that surround that ecosystem but also by destroying the habitats on whose existence depend many sentient and living beings that are disappearing. Just as we cannot separate human wellbeing from the wellbeing of many species on which our life depends, many species are losing their habitats on whose existence they depend for survival, but this time that stable equilibrium has not been disrupted by a natural disaster or a meteorite from outer space. We, humans, have changed and are changing the conditions of life on Earth and upsetting the stable equilibrium of that living ecosystem and many species going extinct. The love of life is what moves people that *care about the wellbeing of that ecosystem that has granted the miracle of life to all living beings*. And the contempt for life is the contempt for the wellbeing of that ecosystem that moves the human species to disregard the *need* of life by threating many species that are suffering and dying from reinforcing that countervalue in the human world. That countervalue is a form of suffering, indeed, but there are many ways to satisfy our need that do not upset the stable equilibrium of that ecosystem on which the miracle of life depends. Find ways to fight need that do not harm the wellbeing of that ecosystem is something that depends on knowing how not to inflict suffering and causing the death of more species. In other words, the realization of the value of wellbeing by human systems is related to realizing the value of knowledge.

How do humans realize the *value of wellbeing* ? Well, it starts with the satisfaction of human need worldwide. Indeed, unlike the realization of the value of knowing by, we cannot realize the value of wellbeing without taking into account the whole planet because that ecosystem is an entangled web of species distributed in different habitats that are not separate from each other as they share the same planet Earth together. The loss of one habitat affects the wellbeing of other entangled species and the wellbeing of the human beings whose lives depend on them. So, as long as there are humans that cannot satisfy their need because they are losing the habitats on which they depend, the value of wellbeing will not be fully realized. Of course, humans are already doing something to prevent the loss of endangered species and the destruction of habitats that host the miracle of life. But the satisfaction of human need does not come from those protected areas but from man-made habitats that feed and shelter most of the world population that is concentrated in cities and that buys its goods from markets that sell anything that has a demand. So to realize the value of wellbeing we will have to replace the entangled sectorial value chains that satisfy our need with others that do not harm the entangled web of species on which the wellbeing of humans depends. The creation of those new industries will come from creating new markets for new products that contribute to the wellbeing of humans *and* that of the ecosystem that has granted us the miracle of life. Companies and buyers cannot continue being moved by the contempt for life when the knowledge that they need to create a different economy to realize the value of wellbeing is already available to fuel a new demand moved by people's love of life. Here we could say, following Socrates, that nobody moved by the love of life would harm the wellbeing of the web of life knowingly but only out of ignorance.

Talking about the relationships between values, the love of life includes taking care of people's health and fighting *illness* which is another countervalue reinforced in the human world through what we pursue. Illness is different from need because they are related to different values. In fact, it could seem that the value to which illness is related, the *value of health*, is a need in the same way as education might appear so, but then everything we love would be a need. However, we are only concerned about our health when we suffer an illness or injury or when we do not take care of it and see signs of deterioration. Likewise, the desire to know is not a need because as I showed when discussing Plato's concept of love as desire, a value is not something that benefits us exclusively but something that helps others. In the case of health and wellbeing, people want to save others from illness and help for those in need out of the love of life. Both need and illness are different forms of suffering that we find in the human world and that are reinforced through what humans pursue and can lead to death if they do not fight them through the realization of the values of wellbeing and health. It goes without saying that the realization of the value of health by human systems also depends on the realization of the value of knowledge and fallibility, as healthcare systems make use of scientific discoveries and medical innovations to diagnose and treat patients that suffer illnesses and injuries.

I cannot go into the different forms of illness as this is not a book on medical science, but I have to say something that concerns the way in which we classify diseases because bearing in mind the anatomy of the universe could be helpful. We know that diseases can affect plants and animals, that is, biological and sentient beings, indeed, but, when it comes to human beings, we have to get rid of the mind-body dualism because, otherwise, everything that is not a physical disease (body) must be a mental disorder (mind). As I said earlier, the well-being of plants depends on the wellbeing of the habitats in which they live and, like animals and humans, they are susceptible to diseases and death. This form of suffering is different from the need and illness that affects animals because, besides the pain that comes from physical disease, they can also experience mental disorders. However, there are other forms of suffering that are neither physical diseases nor mental disorders and that only affect human beings, such as loneliness and meaninglessness, which should not be classified as an illnesses. The psychiatrist that came closest to understanding this in the case of the lack of meaning was Victor Frankl ([1964] 2010) though he could not avoid both relating it to the concept of consciousness or "the dimension of spirit" and describing it in a way that resembles a need, an unfulfilled "will to meaning". He called that genuinely human dimension, the "noological dimension" which was different from "the biological and psychological dimensions".

> Man passes this dimension whenever he is reflecting upon himself – or rejecting himself; whenever he is making himself an object – or making objections to himself; whenever he displays his being conscious of himself – or whenever he exhibits his being conscientious.
>
> (Frankl [1964] 2010: 73)

The author of *Man's Search for Meaning* was on to something but was unable to work his way out of the mind-body dualism and considered that meaninglessness, which he also termed "existential vacuum" or "existential frustration", can result in "noogenic neuroses", "rooted in collisions between different values, or in the unrewarded longing and groping of man for that hierarchically highest value" (Frankl [1960] 2010: 97). Of course, this reminds us of Scheler and Hartmann's work on ethics which Frankl had studied. He also struggled to differentiate the "will to meaning" from the "will to pleasure" because he wanted to avoid conceiving meaning as a need, a term he resisted to use as it could be assimilated to a "drive", but he finally said the word in one of his last lectures.

> Our industrialized society is out to satisfy all needs, and our consumer society is even out to create needs in order to satisfy them; but the most human of all human needs – the need to see a meaning in one's life – remains unsatisfied. People may have enough to live by; but more often than not they do not have anything to live for.
>
> (Frankl [1986] 2010: 203)

But it did not occur to him that *meaning was a value* whose countervalue is meaninglessness or, better, *the absurd*, to use Albert Camus' ([1942] 1955) term. So as not to run ahead of myself, let me continue with *health*. Human systems realize the value of health diagnosing and treating illnesses and injuries to avoid suffering and death through all the medical specialties that humans have developed through learning how to fight diseases and recover our health. The love of life of healthcare professionals goes hand in hand with the knowledge they adquire because the better they become at diagnosing and treating illnesses and injuries, the better they become at realizing the value of health. The people that pursue the healthcare professional vocation have contempt for the suffering and death of those people who they can be through their medical and nursing practice regardless of whether they can pay for it or not. Therefore, we cannot realize the value of health in the human world by providing health only to those who can afford it, just like we cannot realize the value of wellbeing without making poverty history. But, again, other contervalues seem to be at play here, such as greed and cruelty, which is why some countries that could afford it deny the access to healthcare to those who do not have private health insurance or cannot pay for medical prescriptions and bills. It goes without saying that the healthcare systems who realize the value of health cannot afford to see human suffering and death around them because they are paying lip service to the Hippocratic Oath: "I will do no harm or injustice to them". And yet, denying them access to healthcare is doing them injustice. The miracle of life has been granted to everyone and nobody can decide who deserves to suffer and die, which applies equally to the value of wellbeing because nobody deserves to be poor. If we do whatever we can to save the lives of our loved ones, including our pets, why do we treat with contempt the lives of humans who do not deserve to suffer and die? It is certainly not out of the love of life that we do injustice to them.

Our next countervalues is *loneliness*, which seems to be the least understood of all the forms of suffering. As with the current classification of diseases, what is preventing us from understanding this genuine form of *human suffering* is also the mind-body dualism and the lack of a general theory of cybernetic systems to understand what this suffering is about. Loneliness results from losing something that is a prerequisite and defines cybernetic systems. The general theory of cybernetic systems provides a scientific explanation as to why animals are able to maintain a stable equilibrium with their environment (Pretel-Wilson 2020: Annex II). The theory of evolution is able to explain inborn reactions but not variable reactions that are "developed in each individual according to the environment" (Ashby 1947: 44 in Pretel-Wilson 2020). Unlike biological systems, in sentient beings, the environment is part of the system because the organism and its environment are functionally related through a feedback mechanism that maintains the stable equilibrium of the cybernetic system. I will not go into the details of that general theory that explains adaptive behaviour in the animal kingdom. Suffice it to say that in the case of human systems, they have not lost the stable equilibrium with their environment when they suffer loneliness because they are able to function physiologically, but what they have lost is the system, the human system which they shared with other human beings. *People that suffer from loneliness suffer from losing the human system in which they were at home*. Indeed, they are surrounded by many human systems around them but they have lost that human system that was home for them. Being at home is not about having a shelter or a house to live in but about having an attachment to a human system. Another word for loneliness is detachment or uprootedness because the people who suffer it have lost what grounded them to the human world. And this form of human suffering comes from *losing grounding relationships with their loved ones and with the people with whom they share a bond*. I could mention a couple of syndromes that are not clinical diagnoses which are related to loneliness such as the empty nest syndrome and the Ulysses syndrome.

Like all countervalues, loneliness is related to a value, namely, the *value of home*, and is realized by people who develop a strong bond with others which is what keeps families together. However, that attachment between human beings that share a human system is not something exclusive of families and does not even happen in all families. Someone can be at home in other human systems through sharing a purpose that brings them together with other people on a regular basis. That purpose is what bonds people together and what nurtures their attachment which is why people can suffer countervalue after retirement or a divorce and even as a consequence of bereavement or a loss of employment, not to mention the change of residency and moving to another country. Therefore, anything that contributes to losing the ground with the human world reinforce loneliness in the human world and anything that is nurtures the attachment among humans realizes the value of home. Surely, there are many things that human beings and human systems are pursuing that are reinforcing loneliness in the human world and there are many things that human systems can do to fight that epidemic. There is something very wrong going on in our world if loneliness is increasing rather than being kept within normal rates. One cannot help wondering if that is not a symptom of the destruction of our world because it is a clear sign that the direction we are heading towards is increasing our suffering. Maybe we should start thinking about what we can do to prevent that preventable level of human suffering that is reinforced through what human beings and human systems are pursuing in the human world.

Finally, human systems realize the *value of meaning* pursuing something worthwhile. This is the most difficult value to realize because human systems have to do something for others that makes ours a better world. Indeed, this is what the meaning of life is all about: *pursuing good purposes on a regular basis*. We cannot do something good on weekends hoping that someone will grant us the eternal salvation while the rest of the week we carry on doing business as usual. The miracle of life is not granted twice, so we have no option but to take our *only chance to pursue something good out of this unrepeatable opportunity*. And pursuing something good will necessarily involve other people who share the same purpose. Furthermore, that unrepeatable opportunity that humans have been granted by life has to take into account that our purpose in life is not something granted to us by destiny. On the contrary, we have to figure out our own purposes continuously and chances are that we will find worthwhile purposes that are shared by other people who will also have to figure out their own purpose in life.

Finding no meaning in life is another genuine form of human suffering. Again, like loneliness, it is not a clinical condition though

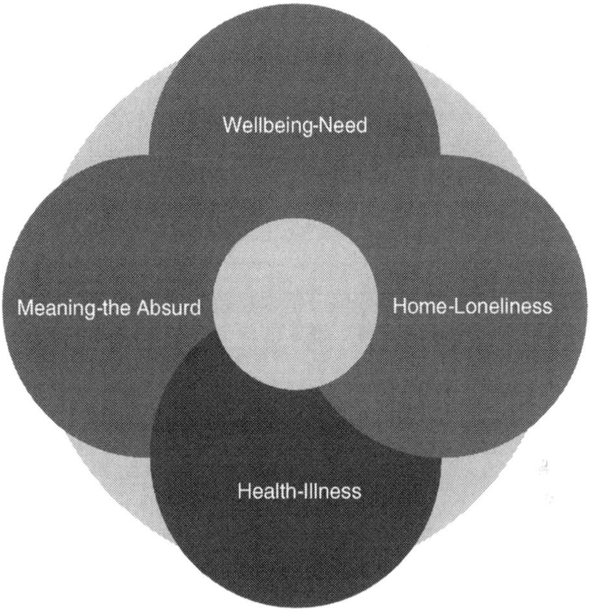

Figure 7.1 The Second Fundamental Law of Ethics: Life-Suffering.

it can end in depression and suicide and, hence, it reinforces a form of suffering that could lead to death.

> More and more a psychiatrist today is confronted with a new type of patient, a new class of neurosis, a new kind of suffering the most remarkable characteristic of which is the fact that it does not represent a disease in the proper sense of the term.
>
> (Frankl [1962] 2010: 49)

We can conclude this chapter with the set of values and countervalues that constitute the second fundamental law of ethics (Figure 7.1).

References

Aristotle (1910). *Nicomachean Ethics*, trans. G. H. Lewes. London: George Routledge.

Camus, A. (1955). *The Myth of Sisyphus and Other Essays*. New York: Alfred A. Knopf.

Frankl, V. (2010). *The Feeling of Meaninglessness: A Challenge to Psychotherapy and Philosophy.* Milwaukee, Wisconsin: Marquette University Press.

Hartmann, N. (2011. *Ética.* Madrid, Ediciones Encuentro.

Hume, D. (2000). A *Treatise of Human Nature.* Oxford: Oxford University Press.

Kant, I. (1988). *Lecciones de Ética.* Barcelona: Editorial Crítica.

Nietzsche, F. (1917). *Beyond Good and Evil.* New York: Modern Library.

Plato (1970). *Republic*, trans. A.D. Lindsay. London: Heron Books.

Pretel-Wilson, M. (2020). *Utopics: The Unification of Human Science, Annex II.* Cham: Springer.

Scheler, M. (1973). *Formalism in Ethics and Non-Formal Ethics of Values*: A New Attempt *toward the Foundation of an Ethical Personalism.* Evanston: Northwestern University Press.

Schopenhauer, A. (1909). *The World as Will and Idea*, Vol. II. London: Kegan Paul, Trench, Trübner.

Schopenhauer, A. (2009). *Parerga y Paralipómena*, Vol. II. Madrid: Editorial Trotta.

Spinoza, B. (1970). *Ethics*, trans, A. Boyle. London: Heron Books.

8 The Value of Fellowship

In this chapter we will encounter another fundamental law of ethics which, unlike the former ones, there are few moral philosophers that have dealt with any of its essential components. On the contrary, most of what we can learn from them is related to countervalues. The only available clue is given by the Scottish moral philosophers who spoke of the "love of our own nation" to distinguish it from the love of humanity, even if it was to deny the latter. But before we start unveiling that value through its corresponding countervalue, let me say something about *patriotism* to avoid confusions. There is a tendency to distinguish patriotism, the duty to serve and flight for one's country, from something more recent that emerged in the 19th century which historians have named *nationalism*. Patriotism is present in Plato's *Republic* in the duty of the warriors to protect the city and flight intruders with courage or valour, which is the virtue most suited to the irascible part of the soul. Aristotle (1910) also mentions courage as the middle term between two vices, cowardice and recklessness, and claims that "the good man [...] does many things for the sake of his friends and his country, even to the extent of dying for them, if need be" (Aristotle 1910: 266). Nationalism, on the other hand, was born out of the Romantic period in Europe beginning at the end of the 18th century with the awakening of the national identity or consciousness of peoples. However, the countervalue of the value we are searching for is not a political ideology of a particular period in human history but something reinforced in the human world from time immemorial.

For the lack of a better word, I will use the same term to describe this countervalue which should not be confused with "the belief in the value of belonging to a group or a culture", which Isaiah Berlin

DOI: 10.4324/9781003649496-10

(1976, 1991) attributes to Herder whom he claims to be the founder of nationalism. "Herder virtually invented the idea of belonging. He believed that just as people need to eat and drink, to have security and freedom of movement, so too they need to belong to a group". Well, I will suggest that nationalism as an immemorial countervalue was present in the work of Herder but especially in Hegel, though it can be argued that it was anticipated by Hume and Smith's through their concepts of sympathy based on proximity. However, they never associated sympathy with the feeling of pride or self-love that is indeed coupled to nationalism. The correlation of values-countervalues with emotions and feelings, which I will treat in the last chapter, is indispensable to gain a better understanding of the fundamental laws of ethics, and absolutely necessary to distinguish nationalism from the *value of fellowship*. Nationalism is coupled to the emotion of envy in addition to the feeling of superiority towards other people. Although a good start, any definition of a countervalue has to be underpinned by the Love-Contempt duality constituting the moral sphere of the human world. I will get to that after introducing what the Scottish moral philosophers had to say about the "love of our own nation" that predated the *contempt for other peoples* found in some German philosophers who justified nationalism.

Well, Smith (1982), probably assuming Aristotle's division between ethics and politics, argued that there is a division between the virtue of prudence, related to egoism, and "our benevolent affections", related to the "virtues of justice and beneficence; of which, the one restrains us from hurting, the other prompts us to promote that happiness" (Smith 1982: 262). The good feelings towards others, however, had no limit because it "may embrace the immensity of the universe" because our imagination cannot exclude "any innocent and sensible beings whose happiness we should not desire". That "universal benevolence", which makes our interests coincide with the interests of others, is less omniscient than it seems, however, despite his claim that "the inferior interests should be sacrificed to the greater interest of the universe, to the interest of the great society of all sensible and intelligent beings" (Smith 1982: 235). That universal benevolence is limited by a hierarchy of proximity which puts first the "Individuals [whose care] are recommended by Nature" followed by "the order in which Societies are by nature recommended to our Beneficence" (Smith 1982: 219–234). Basically, it was a hierarchy of duties because it starts with what requires the most attention, *self-care*, which includes our family, friends, neighbours, and acquaintances, and then comes what becomes the last of our duties, *other-care*, that is, the care for others who are not close to us. But contrary to what we might think, judging from Smith's universal benevolence, other-care does not include other-peoples, because our duty is to "the State or sovereignty in which we have been born and educated". Why? Because, according to him, it is "the greatest society upon whose happiness or misery, our good or bad conduct can have much influence". In other words, neighbouring nations came last because we have a small or no influence on their happiness or misery.

Did this imply indifference towards the fate of other peoples living other countries? We can only assume that our sympathy towards other nations, on which our benevolent affections are grounded, was more remote than our sympathy towards our fellow people. It is more explicit in the case of Hume ([1739] 2000), who associated sympathy to the laws of association of the mind, suggesting that the "principle of sympathy or communication" was the result of converting the idea "of the affection of others" into an impression "by the force of imagination". Unlike Smith, he grounded the principle of sympathy on utility or what "is useful or pernicious to society", which, of course, was different from Smith's concept of sympathy based on feeling the happiness or misery of others as our own, but more coherent with the idea that we feel sympathy towards the people who we are closest to us. "The sentiments of others have little influence, when far removed from us, and require the relation of contiguity, to make them communicate themselves entirety" (Hume [1739] 2000: 207). Indeed, "sympathy depends on the relation of objects to ourselves". Accordingly, the people that are not related to us because they live in different nations are not our concern. However, Hume found a trick to avoid that conclusion without giving up on the application of the law of contiguity – which he had discovered in the first book of his *Treatise of Human Nature* – to morality.

> An *Englishman* in *Italy* is a friend: A *European* in *China*; and perhaps a man would be beloved as such, were we to meet him in the moon. But this proceeds only from the relation to ourselves; which in these cases gathers force by being confined to a few persons.
>
> (Hume [1739] 2000: 310)

With that example, he was also warning us against the idea of a supposed "love of mankind" because there cannot be such a passion "independent of personal qualities, of services, or of relation to ourself". In other words, any object of sympathy needs to be in proximity to us to care about it, thus, for objects far removed from us, "our sympathy is proportionally weaker, and our praise or blame fainter and more doubtful", which suggest that we are indifferent towards other nations that are not related to our own. As we know, contempt implies not only indifference, but also hate, so we are beginning to see that nationalism is related to that indifference towards other nations. However, what neither Smith nor Hume anticipated was that the *contempt for other peoples* was directed not only to the people that live in other territories, but also to peoples from other nations that live in the same country and even towards peoples that share the same nation. However, we cannot say that they hated other people as both moral philosophers were compassionate human beings. In fact, Hume extended sympathy "beyond our own species", because "there is no human, and indeed no sensible, creature, whose happiness or misery does not, in some measure, affect us, when brought near to us" (Hume [1739] 2000: 309). And Smith (1759) did not want to deny the sympathy to other nations and would recommend that "each nation ought, not only to endeavour itself to excel, but from the love of mankind, to promote, instead of obstructing the excellence of its neighbours" (Smith [1759] 1982: 229). But not because the "love of mankind" is more fundamental, "we do not love our country merely as a part of the great society of mankind: we love it for its own sake, and independently of any such consideration" (Smith [1759] 1982).

But values, like countervalues, are mutually related. If they could not see farther, it was because they did not think in terms of human systems nor of different levels of the human world because the value we are searching for is related to peoples and applies to the national level. Indeed, the *value of fellowship* is realized by people out of love fortheir fellow people which must not be confused with *the devotion to one's own nation* that excludes the love of humanity out of the contempt for other peoples. How can I claim so, is not devotion the same as love? Well, with the aid of the correlation between values-countervalues and emotions and feelings, we can understand that that devotion, given that it is coupled to envy and self-love, can turn into a *contempt for other peoples*. Another way to look at it, as we know from questioning Plato's concept of love, is to realize that love is not about having something but about contributing to something we love. In fact, a nation cannot be owned because it is not a human system to which human beings belong but a level of the human world. It would be more appropriate to say that peoples share a nation in the same way as nations share a civilization. That willingness to contribute to a country implied in the love of fellow people is what we see in the people who are willing to fight for their country if they have to. "The patriot who lays down his life for the safety, or even for the vain-glory of his society", says Smith (1759). However, it is not national pride, but anger coupled to the love for their fellow people what moves them. Therefore, it is not for the second reason that Smith suggests, "Upon account of our own connection with it, its prosperity and glory seem to reflect some sort of honour upon ourselves" (Smith [1759] 1982: 227). On the contrary, unlike nationalism, what moves people who realize the value of fellowship is coupled to the feeling of humility.

Well, as I said at the beginning, the work of our predecessors is not very illuminating as regards the essence of the value of fellowship, but they have shed some light on its corresponding countervalue in suggesting that its essence is coupled to the *feeling of superiority*, namely, pride-self-love or arrogance. And the first to do so was the alleged father of nationalism, Johan Gottfried Herder. That superior race was the European, and we can find many instances of it in his *Outlines of a Philosophy of the History of Man* (1784–1791). He traced back the origins of the European civilization to the "Mosaic" tradition, that is, to a monotheistic religion that put the history of mankind on track towards the cultivation of reason as its destiny.

> We of the North, too, have received our sciences in no other way but under the garb of religion, so that we may boldly affirm, from the history of all nations, the earth is indebted for the seeds of all superior degrees of cultivation to religious traditions, oral or written.
>
> (Herder [1784–1791] 1800: 233–234)

After that, our ancestors were able to surpass "the most ancient kingdoms" that had developed in Asia, ruled by despotism and inhabited by tamed people who were used to been oppressed, thanks to "the era of the Grecian republics [which] was the first step of the human mind towards mankind". Indeed, Athens and Sparta solved the question as to "how men should govern men" and we are also "indebted to the Greeks for the basis of almost all our sciences". Similarly, we owe Rome the fate of humanity because at one stage this city "possessed the sovereignty of the world".

But it was later when the European superiority truly took off with the cultivation of reason, fulfilling the destiny of mankind as "superior men". Europe was destined to surpass every other civilization thanks to "a reason and humanity that in time should embrace the world". How did Europe become superior to other peoples? Through "discoveries which, carried into practice, became powerful implements in the hands of posterity", "above all its peculiar industry in the arts, the result of many common exertions, procured it this rank". In short, the "cultivation through the means of industry, arts, and sciences" gave Europe the destiny of humanity. That feeling of superiority is the essence of nationalism because the people who are moved by this countervalue have contempt for other peoples. In contrast, the love for our fellow people is coupled to a feeling opposed to pride, namely, humility.

Herder believed that the European civilization was superior to other peoples because it had excelled in the sciences and the arts, whereas Hegel [1822–1830] (2001) attributed the superiority of the Germanic people to an internal development of the German spirit thanks to Protestant Reformation. Of course, Hegel could not attribute that superiority to something that had originated elsewhere and, thus, he had to find something that made the German people stand out from the rest of Europe. Either way, neither the external nor the internal developments make some peoples better than others. On the contrary, having contempt for other peoples only makes the human world a worse place by reinforcing the aversion towards other peoples with whom we share the same world together. Just as nationalism is defined by a peoples belief in their superiority coupled to the feeling of self-love or pride, it is equally characterized by an "inclination for domination", that Kant talked about in his *Anthropology* (1798), which is related to the emotion of envy. Unlike Hegel, we do not find any allusion to that inclination in Herder's work, which, according to Kant, comes from the "fear of being dominated by others: it tries to avert this by getting a head start and dominating them" (Kant [1798] 1974: 140). However, it is not out of fear that some peoples have colonized other lands to avoid being dominated by other peoples. Instead, the emotion of fear coupled to the love for our fellow people is what counteracts that nationalism coupled to envy that drives peoples who want to dominate other peoples because of what they possess. The Scramble of Africa driven by Western European nation-states that would eventually draw them to the First World War (1914–1918) is a reminder of what nationalism is capable of.

Fortunately, the love for our fellow people is what moves people to realize the value of fellowship through all sorts of purposes that contribute to the realization of the values that constitute the third fundamental law of ethics. We are not completely in the dark when it comes to discovering that group of values but, again, it comes from illuminating some of the corresponding countervalues, though not many, because, again, there was no concept of human system and philosophers could not see that the recipient and custodian of those values was the human world. In fact, the first countervalue I am going to discuss can be found in the most admired exponent of liberalism after Locke, John Stuart Mill, the author of *On Liberty* ([1859] 1864), who also wrote *Considerations of Representative Government* ([1861] 1873), and who, as we will see, was also an exponent of *imperialism*. Equally, the last countervalue I am going to discuss is epitomized by Jean Jacques Rousseau, who is credited for having inspiring the French Revolution, but is also the "father" of *philistinism*, a countervalue that reinforces the contempt for something that completes the third fundamental law of ethics.

If we think in terms of people who share a territory, the first driver is to build a country with its own political institutions for people to direct their own destiny. As we know from the concept of human systems, humans are directed beings that *go towards something from themselves* and, to do so, they need to pursue their own purposes to achieve something. Indeed, humans share a land in which they do things together with other humans, and, in so doing, they pursue their purposes. What is unique about humans that share a land it that the political institutions they share together do not belong to anyone in particular to the country they belong to. There are human beings working together and sharing the purpose of those political institutions, but those human systems themselves are not owned by the people who work there carrying out their public duties. These political institutions are so fundamental to the realization of the *value of freedom* that without them people would not be able to pursue their own destiny. The love for our fellow people is what moves people to realize the value of freedom through the establishment of political institutions.

I will not enter as to which is the best form of government for a nation, all I can say is that it cannot be hijacked by humans who want to pursue their own agendas through the political institutions of the nation. If the people who share a territory cannot go towards something from themselves, then they cannot pursue their own destiny as directed beings entangled together with other people. Any form of government that does not serve its people cannot realize the value of freedom because it will not let them pursue what they want to achieve with their lives. The people need to be able to go towards something from themselves which means that they cannot be directed by others who are not moved by the love for their fellow people and, hence, do not care about them. Many political philosophers have talked about the different forms of government, from Plato, who wanted the city to be ruled by a philosopher king, to Locke, who wanted a legislative power to end the discretionary power of an absolute monarch. All peoples will have to find the political institutions that best serve them, trying to avoid falling into opposite extremes, namely, enlightened despots who know what is best for their fellow people and democracies in which fellow people vote for rulers who do not care about them. What is most important is that the legislative and the executive powers are separated because the people whot rule cannot be the same people who pass the laws. Indeed, the system of direction of a state has to distinguish both functions to serve its people. Likewise, there has to be a representative of the government and a representative of the parliament, which cannot be the same people and which cannot hold those positions indefinitely because they do not own the political institutions of the nation and are only there to serve its peoples.

Maybe I should distinguish the value of freedom from the concept of freedom of the fathers of liberalism which has its roots in Hobbes, who, by the way, was an advocate of absolute power which I have warned against on the grounds that this form of government cannot realize the value of freedom in the human world. For the author of the *Leviathan* (1651), "LIBERTY, or freedom, signifieth properly the absence of opposition (by opposition, I mean external impediments of motion)" that applies to equally to everything in the universe, "no less to irrational and inanimate creatures than to rational" (Hobbes [1651] 1909: 161). In the case of human beings it meant, "the absence of external Impediments: which Impediments, may often take away part of a man's power to do what he would". Similarly, for Locke (1689), "*liberty* is, to be free from restraints and violence from others", adding, "which cannot be, where there is no law". And that liberty is the "*liberty* to dispose and order as he lists, his person, actions, possessions, and this whole property" (Locke [1689] 1821: 234) Unlike the "*freedom of nature*", which is "under no other restraint but the law of nature". Therefore, with Locke, liberty becomes the freedom of the members of a "civil society" who have decided voluntarily "to join and unite into a community, for their comfortable, safe and peaceable living one amongst another". Finally, for the author Mill ([1859] 1864), liberty is the "social or civil liberty" of an individual within society that has to be protected from the illegitimate intervention of state and from the "tyranny of the majority", and "The only purpose for which power can be rightfully exercised over any member of a civilized community, against his will, is to prevent harm to others" (Mill [1859] 1864: 22).

This concept of freedom implies that there is an original contract between the members of a community that ensures their pursuit of purposes within a state as long as that pursuit does not harm others, who are also entitled to pursue their own purposes without harming anybody. In the *United States Declaration of Independence* (1776), it stands as "the pursuit of Happiness". Of course, that pursuit had to be specified further because people do not have the right to everything to be happy, so the concept of liberty came to be defined by the concept of *rights and freedoms* of human beings that we find in the "bill of rights" or the "fundamental rights" in constitutions and in the *Universal Declaration of Human Rights* (1948), all inspired by the *French Declaration of the Rights of Man and of the Citizen* (1789). These are foundational laws realized by political institutions that underpin the legal sphere of many nations and some civilizations, but I am trying to get to a value that drives peoples throughout the entire human world. The value of freedom I am speaking of is the *freedom of people to pursue their own destiny*. The history of our world is full of people who have fought and are fighting for their freedom and to put some flesh into their stories you can read *Freedom in Exile (1991)*, an autobiography of the 14th Dalai Lama, and *Long Walk to Freedom* (1994), another autobiography of the first president of South Africa after the end of apartheid regime, Nelson Mandela.

But the freedom of people means the freedom of the humans that live in that country and, therefore, for the value of freedom to be realized, they have to be able to *pursue their own purposes*. In fact, strictly speaking, human systems realize the value of freedom within human systems in which human beings can be free but only if they are able to pursue purposes. However, the fact that people can pursue something within a human system does not necessarily mean that they are free because they would rather do something else if they could. In other words, people may not be doing what they would like to pursue in life if they had a choice. Not everyone is happy in what they do for a living but they have to continue working and doing the things they do not like because they need the money to pay the bills and cannot risk losing their income unless they find something better. Some people might be lucky enough to find one day a job in a human system in which they can pursue what they like together with other people who are also fortunate. Therefore, the creation of new human systems can contribute to the realization of the value of freedom in the human world by creating jobs in which people want to invest their lives. Human systems that provide jobs are not the only ones that can realize the value of freedom, as other human systems, such as families, can also help human beings pursue their purposes by supporting them in their aspirations in life.

It is time to introduce the contervalue of the value of freedom, *imperialism*, through Mill's *Considerations of Representative Government* ([1861] 1873) where it is justified. Prior in *On Liberty* ([1859] 1864) he had already said something relevant by way of an introduction to whom the concept of freedom applies, "only to human beings in the maturity of their faculties". "For the same reason, we may leave out of consideration those backward states of society in which the race itself may be considered as in its nonage" (Mill [1859] 1864: 23). Before people are prepared to be free, "Despotism is a legitimate mode of government in dealing with barbarians, provided the end be their improvement, and the means justified by actually effecting that end" (Mill [1859] 1864). Of course, there are also enlightened rulers, such as "an Akbar or a Charlemagne", who are preferable for those people to obey. A couple of years later (1861), he would expand on this point saying that in order to reap "the benefits of civilized society", "a civilized government, to be really advantageous to them, will require to be in a considerable degree despotic" (Mill [1861] 1873: 13). That civilized government, however, is no longer the ruler at the helm of the political institutions of the nation, but a foreign government and "the people for whom the form of government is intended must be willing to accept it". Indeed, the obedience to a native ruler is now replaced by the docility of the people under a foreign ruler who knows what is best for them. "In politics as in mechanics, the power which is to keep the engine going must be sought for *outside* the machinery" (Mill [1861] 1873: 21). And yet, that civilized government is a blessing for those "rude people" who would not be able to climb the ladder of civilization otherwise, "a people in a state of savage independence [...] is practically incapable of making any progress in civilization until it has learned to obey" (Mill [1861] 1873: 47). He even goes as far as saying that a slave, "who has not learned to help himself", is ahead of the savage because he has learned that lesson. However, a civilized government should spare them that trouble, "A people of savages should be taught obedience, but not in such a manner as to convert them into a people of slaves" (Mill [1861] 1873: 51).

This is the first form of nationalism which consists in the *subjugation of other peoples* by foreign people who reinforce the countervalue of imperialism in the human world through what they pursue. And if someone is wondering how imperialism is related to the feeling of superiority, wait and see what Mill has to say about it. A foreign government can "cure or alleviate the evil" of the people who suffer from "extreme passiveness, and ready submission to tyranny", but only if that government comes from "a superior people or a more advanced state of society" that can "carry[...] them rapidly through several stages of progress" (Mill [1861] 1873: 91). In other words, there are foreign rulers "superior in civilization" that can help them get out of their inferior state of society. Furthermore, inferior people will greatly benefit from being absorbed by superior people as there is "a gain to civilization". Conversely, "The absorption of Greece by Macedonia was one of the greatest misfortunes which ever happened to the world". However, if the superior is smaller in number, "the conquerors and the conquered cannot in this case live together under the same free institutions. The absorption of the conquerors in the less advanced people would be an evil" (Mill [1861] 1873: 316). As to who should govern the colonies, if the they are of "similar civilization to the ruling country", then they are entitled to have a "representative government", "such as the British possessions in America and Australia", If not, "like India", they "must be governed by the dominant country, or by persons delegated for that purpose by it", that is, by the British government, which "most facilitates their transition to a higher stage of improvement", and it should appoint a "British minister" who "should be responsible to the British Parliament". The only case in which Mill favoured a people to be governed by themselves is when the choice was between native or foreign despotism. Better to be subdued by their own despotic rulers given that "Foreigners do not feel with the people" (Mill [1861] 1873: 348).

The next countervalue related to nationalism is *racism* and consists in a people's *contempt for other peoples*, which can either mean the rejection of foreign people who settle in "their" country or of native people with whom they do not want to share the same land. One is the rejection of people who have left their homeland to live in another country and the other of indigenous people who already live in their homeland but, in both cases, that rejection is driven by people's feeling of superiority and their hate other peoples that reinforces racism in the human world. I am not going to go into what philosophers have said about other peoples because it would be derogatory, suffice is to say that racism has been justified on many grounds related to the superiority of one people over others. However, immigration and hybridization have always enriched a people and rejecting other peoples is also rejecting their own ancestry. There are no pure races and no pure ethnic groups because interbreeding is healthy and makes people more diverse. Not even our species is pure as we carry DNA from the Neanderthals and older species of humans also interbred with them, so all humans share a common ancestry and, thus, we cannot say the Homo sapiens are superior to other humans that are now extinct of which we still carry their DNA. So the superiority of a race cannot be justified scientifically. We do not know enough about why the Neanderthals became extinct but it is more than probable that they might have lived together with our species as fellow species. In any case, the countervalue of racism is related to a corresponding value which is already suggesting itself, the *value of equality*.

The value of equality can only be realized by people who grant the same *opportunity to pursue their own destiny* to other people. Indeed, people can share the same territory or country with other people as fellow people. The national level of the human world is not a hierarchy of peoples because that would prevent the value of equality to be realized. On the contrary, it is the contempt for other peoples what reinforces a pseudo hierarchy between higher and lower peoples, developed and developing nations. Human systems have to grant that *opportunity to everyone,* regardless of their physiological differences and, therefore, realizing the value of equality, means counteracting racism, but also ageism, sexism and ableism, which are different forms of the same countervalue.

So far, we have covered the first half of the essential components of the third fundamental law of ethics. It is now time to turn to another countervalue related to the devotion to one's own nation to such an extent that it leads to the rejection of the people who are different from them. Again, for a lack of a better name, I will term what moves people who reject other people who are not like them *absolutism*, but should not be confused with the absolute power of monarchies in Europe during the 17th and 18th centuries. Absolutism can be defined as the *rejection of fellow people*. It is different from racism because absolutism rejects people who are accepted as fellow people. Indeed, fellow people suffer absolutism because of being different. And yet, people are neither different nor equal by nature, because some people are equal in some ways and different in many others. But people who are moved by absolutism do not accept certain differences and reject the people who do not hide them. That makes fellow people have contempt for themselves because they do not want to be excluded by their own people. It is bad to be rejected by fellow people but it is worse to have contempt for oneself. So absolutism, unlike the other forms of nationalism, is reinforced by people who have contempt for their fellow people who do not conform to the norm. There is an absolute model that everyone has to embrace if they want to be accepted by their fellow people whom they do not want to offend.

Furthermore, this is not a rejection based on something that can harm others and from which fellow people have to be protected. No, these are things that do not harm anybody except the people who are offended by them because they think it is something wrong because other fellow people think so. Yet, those differences not only do not harm anybody but are shared with other fellow people who are also rejected by their own people. If they have them in common with other fellow people and are not harmful to others, then those differences cannot be wrong. However, for the people who are moved by absolutism they are wrong and will do what it takes to make them think that way. The tyranny of the majority will reject them by reinforcing absolutism around them. How? By mean of excluding fellow people who display those common differences and despising them as being "abnormal", even keeping away from them so as not to be infected and, in some countries, treating them as criminals and even getting rid of them. The people who reinforce absolutism make them think there is something wrong with them and to such an extent that they despise themselves and envy the people who are "normal".

The people who suffer absolutism in their own flesh will be identified with what I am saying about their fellow people who have contempt for them because they are different from the rest. Paradoxically, it is fellow people who do not suffer absolutism who can do the most to counteract that countervalue by means of realizing the *value of plurality*. However, they do so not out of pity or compassion but out of love for their fellow people because they care about them. Only when all the differences among fellow people are recognized as *something natural* will the value of plurality be realized in the human world. Like always, among the fellow people who do not suffer absolutism, human systems can do the most to fight the rejection of fellow people through the purposes they pursue that realize the value of plurality. I am not going to name all the forms of exclusion that need to be eradicated from the human world because doing so could mean leaving out some and reinforcing some form of absolutism given that nobody can recognize all the natural differences among fellow people that need to be accepted as such.

I would say that we can judge the moral progress at the national level of the human world by how fellow people treat their own people. And that means asking to what extent have human systems realized the value of plurality at the national level of the human world? Like with the value of wellbeing, the value of fellowship cannot be realized in one part of the world while the rest suffers absolutism, in the same way as one part of the planet cannot live in abundance while the other regions suffer from need. People are not guilty from having been born in a country that does not recognize the differences among fellow people and that tries to persuade and convince them that there is something wrong with them. The only thing wrong is not doing anything about it and letting fellow people suffer absolutism and even their loved ones.

At last, we can now complete the last essential component of the value of fellowship and I will do so, as I have anticipated earlier, by introducing the last form of nationalism through the work of Jean-Jacques Rousseau, the "father" of *philistinism*. This political philosopher had *contempt for fellow intellectuals*, maybe because he had contempt for himself as a man of letters and composer. The author of *The Social Contract* (1762), had won a prize a decade earlier granted by the Academy of Dijon for a discourse on a proposed question: "has the restoration of the arts and sciences had a purifying effect upon morals?". Apparently, he was struck by an "illumination" in 1749 on his way to visit Diderot in prison "when the solution of the problem of human vice and virtue came upon him with a blinding flash of inspiration" (Berlin [1952] 2002). He thought he had discovered the origin of human evil and knew that what he was about to say in the

Discourse of the Arts and Sciences (1750) was so radical that he had to warn the reader.

> I foresee that I shall not readily be forgiven for having taken up the position I have adopted. Setting myself up against all that is nowadays most admired, I can expect no less than a universal outcry against me, nor is the approbation of a few sensible men enough to make me count on that of the public. But I have taken my stand, and I shall be at no pains to please either intellectuals or men of the world.
>
> (Rousseau [1750] 1987: 1)

Indeed, the arts are "external ornaments" of society that are "foreign to virtue". "Before art had moulded our behaviour, and taught our passions to speak an artificial language, our morals were rude but natural" (Rousseau [1750] 1987: 4). All the vices we find in civilized people come from the arts and sciences: "our minds have been corrupted in proportion as the arts and sciences have improved". He was attacking his own Age of Enlightenment which he opposed. But b saying that, however, he was not blaming the sciences per se, but the *arts and letters* instead. He even claims that the Greeks, "who twice vanquished Asia", were able to do so because "Letters, as yet in their infancy, had not corrupted the disposition of its inhabitants" (Rousseau [1750] 1987: 6), but, with "the progress of the arts", they were conquered by the Macedonians. And the same happened to the Eastern Roman Empire, "that refuge of the arts and sciences", which was conquered by the Ottoman Empire. In contrast, there are other peoples, such as "the first inhabitants of Persia" but also the Scythians, the Germans and "even Rome in the early days", that have "be[en] preserved from the contagion of useless knowledge". According to Rousseau, some wise Greeks like Socrates had already warned against "the artists and learned men of his day".

I take the term philistinism from Schopenhauer ([1851] 2009) who described a philistine as "*a man without intellectual needs*".

> The philistine's life is animated by no desire to gain knowledge and insight for their own sake, or to experience that true aesthetic pleasure which is so nearly akin to them. If pleasures of this kind are fashionable, and the philistine finds himself compelled to pay

> attention to them, he will force himself to do so, but he will take as little interest in them as possible.
>
> (Schopenhauer [1851] 2009: 361)

Instead, for philistine "oysters and champagne are the height of existence; the aim of his life is to procure what will contribute to his bodily welfare". Without having to describe the philistine as a hedonist, it did have something that would later be criticised as *utilitarianism* by writers who were concerned about it, such as Charles Dickens. In the first paragraph of the first chapter of *Hard Times* (1854), titled "The One Thing Needful", the British novelist introduces what an education based on the principles of philistinism looks like.

> Now, what I want is, Facts. Teach these boys and girls nothing but Facts. Facts alone are wanted in life. Plant nothing else, and root out everything else. You can only form the minds of reasoning animals upon Facts: nothing else will ever be of any service to them. This is the principle on which I bring up my own children, and this is the principle on which I bring up these children. Stick to Facts, Sir!

This was the model of teaching reinforced by Mr. Gradgrind, the school board Superintendent, in which there was not room for the "idle imagination" but only for "realities", it was a system in which the "fanciful imagination" had to be "strangled", "infants were never to wonder". We know very well what Dickens was talking about since the human being is essentially an utopico-practical being, that is, a creator. And that education system was reinforcing one aspect to the exclusion of the other, at a time when the Industrial Revolution was changing England and needed produce "eminently practical" people, which, of course, was much of a concern for the intellectuals of the Victorian era like Thomas Carlyle, to whom he dedicated the book. The people moved by philistinism have *contempt foreverythingthat is not practical* and for the fellow intellectuals who do not produce anything useful. And, like Mr. Bounderby, a successful business associate of Mr. Gradgrind "who could never sufficiently vaunt himself a self-made man", they need to show their "superior[ity] in point of wealth, or rank, or influence and power to other people, who thereupon pay him honour", according to Schopenhauer. For him, that pursuit of "rank and riches, power and influence" comes from vanity as a way to compensate for what they are lacking internally with external recognition. However, it seems more likely to be coupled to envy, envy towards the fellow people that produce non-practical works, an emotion which the philosopher also mentions claiming that it comes from "an unpleasant sense of inferiority, he experiences, in his heart". But it does not seem to be the case because all the other forms of nationalism are coupled to pride-self-love. So Dickens was right in describing the utilitarian as "A man who was the Bully of humility".

And yet, the "external ornaments" are the arts and literatures *that embellish our world* through what humans do for other fellow people who they care about and, by doing so, they realize the *value of culture* and thus the value of fellowship in the human world. In contrast, what defines philistinism is the contempt for culture and the *cult of technology* that shows no concern for their fellow people besides using them for ill purposes. The people who reinforce philistinism in the human world are very proud of how technology is transforming our world, albeit not always for the better, and are reinforcing something dangerous, the "faith in machinery", to use Matthew Arnold's (1869) expression.

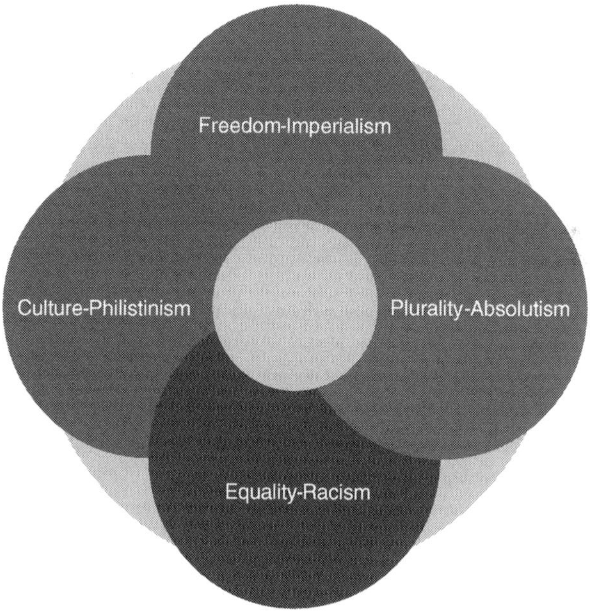

Figure 8.1 The Third Fundamental Law of Ethics: Fellowship-Nationalism.

We can conclude this chapter with the set of values and countervalues that constitute the third fundamental law of ethics (Figure 8.1).

References

Aristotle (1910). *Nicomachean Ethics*, trans. G. H. Lewes. London: George Routledge.

Arnold, M. (1869). *Culture and Anarchy*. London: Smith, Elder and Co.

Berlin, I. (1976). *Vico and Herder*: Two Studies in the History of Ideas. London: Hogarth Press.

Berlin, I. (1991). Two Concepts of Nationalism. *New York Review of Books* (NYRB), 21: 19–23.

Berlin, I. (2002). *Freedom and Its Betrayal*, London: Chatto & Windus.

Dalai Lama XIV (1991). *Freedom in Exile*. London: Abacus.

Hegel, G. W. F. (2001). *The Philosophy of History*. Kitchener, Ontario: Batoche Books.

Herder, J. G. (1800). *Outlines of a Philosophy of the History of Man*. London: L. Hansard.

Hobbes, T. (1909). *Leviathan or The Matter, Forme and Power of a Commonwealth Ecclesiasticall and Civil*. Oxford: Clarendon Press.

Hume, D. (2000). A *Treatise of Human Nature*. Oxford: Oxford University Press.

Kant, I. (1974). *Anthropology for a Pragmatic Point of View*. The Hague: Martinus Nijhoff.

Locke, J. (1821). *Two Treatises of Government*. London: Whitmore and Feen.

Mandela, N. (1994). *Long Walk to Freedom*, London: Little, Brown and Company.

Mill, J. S. (1864). *On Liberty*. London: Longman, Green, Longman, Roberts & Green.

Mill, J. S. (1873). *Considerations on Representative Government*. New York: Henry Holt.

Rousseau, J. (1987). *Basic Political Writings, Discourse on the Sciences and the Arts*. Indiana, Cambridge: Hackett Publishing.

Schopenhauer, A. (2009). *Parerga y Paralipómena*, Vol. I. Madrid: Editorial Trotta.

Smith, A. (1982). *The Theory of Moral Sentiments*. Indiana: Liberty Fund.

9 The Value of Humanity

The last fundamental law of ethics is what unites peoples and civilizations because it applies throughout all the scales of the human world, from the subnational to the supracivilizational levels, and throughout the universe. We do not need to demonstrate that it applies in other human worlds besides our own, in the same way as we do not need to find a place in the universe in which the fundamental laws of physics do not apply to demonstrate their universality. It will remain a hypothesis until proven wrong, and to disprove it we will need something better to replace it with, in the same way as we could not do without Newton's law of universal gravitation until something better could fill the void, namely, Einstein's theory of general relativity. Now that I have mentioned gravity, I should also mention something that has driven physicists crazy since the famous Solvay 1927 Conference, which gathered together the greatest number of Nobel Prize laureates that the world had ever seen together. That is when the foundation of physics was questioned by a "new physics", that is, quantum mechanics, which seemed to contradict what we knew about the laws of the universe. Gravity and the space-time continuum on which it was grounded did not seem to apply because particles behaved in weird ways at the subatomic level. Well, in the case of our "new ethics", we have something similar with what happened when Newton's laws that got rid of Aristotle's cosmology divided into the "sublunary" and "supralunary" spheres by unifying the skies. The four fundamental laws of ethics apply at the subnational *and* supracivilizational levels of the human world.

How do we go about introducing a fundamental law that defies what we know about the human world, a law that questions the

DOI: 10.4324/9781003649496-11

foundation of the "old ethics" grounded on the individual? To be fair to my predecessors, I have to say that they were all well aware that ethics could not be reduced to selfishness because that was not right. There had to be other motives that moved human beings to contribute to society because if ethics was only about pursuing things that pleased ourselves, then moral philosophy would not inform political philosophy, and we would live in a world of *homo homini lupus est* and, to use Hobbes expression, our life would be "solitary, poor, nasty, brutish, and short". Even if it did not occur to them to question the grounding of ethics on the individual human being, they would not accept selfish individuals that did nothing for others. Living together peacefully could not depend on law enforcement and on the fear of being punished for doing injustice to others. It was obvious to them that human beings are not indifferent to others as they care about them, even if they care more about their loved ones and their fellow people. Furthermore, Aristotle (1910) also noticed that people also help those whom are not close to their heart, "for one may goodwill both towards people whom one does not know, and without their knowing it"" (Aristotle 1910: 258). In fact, *benevolence* preceded friendship as "it does seem to be the starting point of a Friendship".

Indeed, benevolence is the starting point we have to follow if we want to do justice to the moral philosophers that preceded us in making sense of the *value of humanity*. Aristotle did not say much about the "feeling of goodwill" other than that it was the origin of friendship, which he was not sure whether it was "a virtue or implies virtue", but it was "most necessary" to be happy. We need affection from people to be happy and, therefore, benevolence was about affectionate feelings towards others. So we will have to start with Spinoza's *Ethics* (1677) to work our way chronologically. We have to remember that Aristotle and Spinoza are separated by the medieval period, so the Greek benevolence is not the Christian kindness. That universal brotherly love was something strange to them because it was only acceptable between equals, that is, between free people. For Spinoza, "Benevolence is the desire of benefiting those whom we pity", which "depends upon education".

> Parents, by reprobating wicked actions and reproving their children on the committal of them, and on the other hand, by persuading to and praising good or right actions, have brought it about that the former should be associated with sadness and the latter with joy.
> (Spinoza [1677] 1970: 134)

In other words, the cornerstone of ethics, caring about others, was built on pity and "those whom we pity" are dictated by others rather than our emotions. I will say more about the emotion of compassion when the time comes, but it is sufficient to say that Spinoza distinguished pity from compassion. "Compassion is love in so far as it so affects man that he rejoices at the good fortune of another and is saddened at his evil fortune" (Spinoza [1677] 1970: 133). And nobody is taught by anyone to feel compassion for humans *and* for animals.

The next moral philosophers to justifying the caring about others were Hume and Smith. I have already mentioned their attempts to establish the field of ethics in Chapter 3, but now we need to understand how their saw the benevolent affections towards others. For starters, according to Hume (1739), "self-love" is an oxymoron because "the *object* of love and hatred is some other person". "Our love and hatred are always directed to some sensible beings external to us" (Hume [1739] 2000: 214). Moreover, morality could not be derived from selfishness.

> So far from thinking, that men have no affection for any thing beyond themselves, I am of opinion, that tho' it be rare to meet with one, who loves any single person better than himself; yet 'tis as rare to meet with one, in whom all the kind affections, taken together, do not over-balance all the selfish.
>
> (Hume [1739] 2000: 313)

Indeed, human beings are capable of "the affections of others", thanks to the principle of sympathy, and can be affectionate beyond those closest to them, showing "*generosity, humanity, compassion, gratitude, friendship, fidelity, zeal, disinterestedness, liberality*, and all those other qualities, which form the character of good and benevolent". Many of those had already been mentioned by earlier moral philosophers and, like Aristotle, he distinguished generosity from liberality. Smith (1759) was also of the same opinion regarding selfishness, as shown in the opening sentence of his book.

> How selfish soever man may be supposed, there are evidently some principles in his nature, which interest him in the fortune of others, and render their happiness necessary to him, though he derives nothing from it except the pleasure of seeing it. Of this kind is pity or compassion, the emotion which we feel for the misery of others.
>
> (Smith [1759] 1982: 9)

That emotion was sympathy which was related to the "social passions" that are "peculiarly agreeable and becoming", such as "Generosity, humanity, kindness, compassion, mutual friendship and esteem, all the social and benevolent affections", and contrasted them with the "unsocial passions" related to "hatred and resentment" that produce "ungraceful and disagreeable" feelings. Again, this list is similar to Hume's, though he distinguished compassion from sympathy, which are the same and not unique to human beings alone because animals also experience that emotion. In any case, the two Scottish philosophers agreed that generosity and friendship are related to benevolence. Yet, one wonders what "humanity" meant for them, because, as we know, sympathy gets weaker the more remote human beings are from us and from our own people, according to them.

With Kant (1775–1781), benevolence gets pushed aside because ethics cannot be grounded on affections or passions and, instead, he speaks of duty, "duty towards his fellows", which is preceded, like in Smith, by "the duty towards himself" and "we can fulfil the former only in so far as we first fulfil the latter". In fact, we can also fulfil the latter without fulfilling the former "by leading a proper life" and still "possess a certain inner worth" or dignity. Accordingly, the caring of others is also covered by the "the moral motive", which are "acts prompted by our love of mankind and by our desire to promote the happiness of the human race". And he contrasts friendly with selfish acts, given that there are only "two motives that move men. The one – self-love – is derived from himself, and the other – the love of humanity – is derived from others" (Kant [1775–1781] 1988: 244). However, Kant grounded the love of mankind on the moral imperative dictated by reason. That moral imperative was so strong that it could even accept to sacrifice our own life to avoid degrading the dignity of humanity in our own person.

By defining the caring about others, including ourselves, as the respect for humanity, Kant ([1785] 2001) was granting an "absolute value" to the person following the categorical imperative dictated by reason, that is, to humans who are not moved by selfish interests and who always treated other human beings as ends in themselves but never as mere means. Furthermore, if the person has an absolute value, they should never risk their lives to save others, as doing so would be equivalent to treating their own person as a means. Kant (1775–1781) would not let the love of others get in the way of the categorical imperative as it could put at risk the duty to ourselves. It is difficult to see how we would help others if doing so could be detrimental to ourselves, and others would do the same to avoid the any negative consequences on themselves. His way out of this paradox was to say we help others in the hope that they will help us when we need them, and since everyone wants the same for themselves, everyone will help each other. "This is the Idea of friendship, in which self-love is superseded by a generous reciprocity of love" (Kant [1775–1781] 1988: 245). But by trying to justify our caring about others that way, Kant was letting in selfishness or self-love through the back door by saying that we help others to help ourselves.

Anyway, the love of humanity is not derived from self-love because *caring about others comes from loving others, not ourselves*. To love others, we have to do something for them showing that we care about them and, as with all the values, the realization of the *value of humanity* comes from realizing a group of values. But before I start introducing the values with their corresponding countervalues, I should say something about the love of humanity to distinguish it from the love of life. The love of life is the love of the opportunity to live that we and other living beings have been granted. In both cases, we love others, but in the case of the love of humanity, *we do something to help others beyond looking after them and making sure they are happy*. This is an important point given that we do not help others out of pity but because we want to help them even if we cannot help feeling compassion for them when when see them suffering. Furthermore, if we only helped others out of pity because we cannot bear their suffering, then we would not be helping them out of the *love for others* but out of the contempt for suffering. When we help others out of the love of life, it is because we cannot bear to see them suffering, but we can help others without having to see them suffering or even to prevent their suffering. Indeed, the love for others is about *preventing human suffering* through what we do for others on a regular basis, which is how human systems realize the value of humanity.

And the countervalue of the value of humanity is *misanthropy*, for whose understanding we can start with Kant. Those who love humanity do something about it because they are concerned about the contempt for others reinforced by the different forms of misanthropy. But what is misanthropy exactly?

> Misanthropy is hatred of mankind and may arise from either of two sources, shyness or enmity. In the first case, the misanthrope

> is afraid of men, deeming them all his enemies; in the second, he is himself the enemy of others. The first is temperamentally timid, thinks himself not good enough for others, underestimates himself and so from very desire for honour flees from men and hides himself. The second shuns other people on principle, considering himself too good for them.
>
> (Kant [1775–1781] 1988: 255)

The first part of his description reminds us of Hobbes's state of nature, but then he links "shyness" and "enmity" to humility and pride, which are human feelings that are not helpful in clarifying what misanthropy is. All we know from what Kant says is that it is related to the "hatred of mankind" because it is coupled with *anger towards others in general*. However, we still need to describe this countervalue to whose reinforcement anger is coupled to. For that, we need to understand that those two sources which Kant mentions are not humility and pride. On the contrary, misanthropy is coupled to guilt. But misanthropy itself is defined by what defines contempt, which, as we know, is *not reducible to hatred alone because it includes indifference towards others or lack of concern for others*. This means that some forms of misanthropy will be related to the hatred and others to indifference. In addition, someone can hate and be indifferent to humanity in general without necessarily having contempt for anyone in particular. Indeed, we can have a lack of concern for others who are not our loved ones or our fellow people.

Caring about our own fellow people is not necessarily the same as loving humanity if we do not care about others besides them. So, how do we care about other human beings? Well, by realizing the *value of generosity* beyond providing money to release their suffering, that is, through what human systems can do for others to prevent their suffering. If we want to separate the value of humanity from pity, we cannot reduce generosity to charity because it could be, in fact, reinforcing misanthropy if we do so to feel better with ourselves. Indeed, we cannot care about others if that help is coupled to pride-self-love. This does not mean we do not have to help others in need. On the contrary, we have to help them with all we can do for them, not just with money. *We need to take generosity seriously because a clear understanding of this value is absolutely essential to prevent the destruction of our world.* So let us see what we can learn from the way in which moral philosophers have treated such an important topic for saving our world.

Let me start with Aristotle. The separation of generosity from *liberality* comes from his understanding that there is a virtue related to wealth or property, namely, "all those things whose worth is measured by money", that consists in a mean between two vices, prodigality and illiberality or stinginess, depending on whether there is an excess or a defect in giving or spending. Even though both are "reprobable" passions, the excess of generosity is preferable to stinginess. Conversely, stinginess, as an excess of receiving, is contrasted to prodigality. Unlike them, "the Liberal are liked almost best of all virtuous characters, because they are profitable to others, and this their profitableness consists in their giving" (Aristotle 1910: 95). "He who spends on right objects and right amounts and at right times" is the virtue of liberality. Thus, the liberal spends and acquires "in right proportion" and, therefore, unlike the illiberal, his business is not "the art of procuring property". Aristotle is alluding to greed. "For the man who is more pleased than he ought to be with every acquisition and more pained than he ought to be at every expenditure is illiberal" (Aristotle 1915: 1231). However, he associates greed with the illiberal who receives more wealth than is necessarily *and* who does not spend enough of it when it is necessary as the vice of illiberality. Greed, in particular, is described as the "pursuit of disgraceful gain, in so far as such men seek gain from all sources and think more of the profit than of the shame". "It belongs to illiberality to value money above everything, and to think no reproach can ever attach to what yields a profit" (Aristotle 1915: 1251). In summary, Aristotle distinguishes generosity from liberality on the grounds that that virtue is related to wealth and, therefore, the "liberal man" avoids spending *and* acquiring more wealth than is required. But should we define generosity in relation to wealth? Have I not said earlier that generosity is not about giving money or part of our wealth to others? Indeed, human systems realize the value of generosity by *helping others through what they do on a regular basis to prevent their suffering.*

Hume (1739) does not say much about generosity besides including it with the "benevolent or softer affections" of which the "good-will of mankind" is capable of and implying that it is close to benevolence. And yet, he also seems to imply that "acts of generosity" come from wealth which takes us back to Aristotle's liberality. Smith (1759), on the other hand, tells us that generosity, like humanity, "are the qualities most useful to others". "Those two qualities, which at first sight seem so nearly allied, do not always belong to the same person. Humanity is the virtue of a woman, generosity of a man" (Smith [1759] 1982: 190). Apparently, humanity means "tenderness", which, by the way, can be found in women and men, and generosity, as expected, is linked to wealth. "That women rarely make considerable donations, is an observation of the civil law". Again, generosity does not depend on gender. But then he says that generosity is different from the other affections linked to sympathy because it requires "self-denial" or sacrifice of our own interests; for instance, giving up something to help others, "some great and important interest", "an office" or even "the man who exposes his life to defend that of his friend", "because they feel more exquisitely what concerns that other person than what concerns themselves" (Smith (1759) [1982]: 191). Interesting, but to repeat it again, generosity is not a spontaneous selfless act but a value that is realized on an ongoing basis. Finally, Kant bundles generosity with "kindness and sympathy" and discusses it in relation to "promoting the welfare of others" and to the Idea of friendship I have already discussed, "the happiness of each would be promoted by the generosity of the others". Then, generosity seems to spring from that strange altruistic self-love or self-interest. No, generosity cannot be derived from a countervalue which is related to a different value.

Despite being such an important value, moral philosophers do not say much more about generosity. Aristotle does not consider it a virtue and for many it is about giving out something to people for whom we feel pity, that is, *charity*. Ironically, in our present time, the countervalue related to generosity is so entrenched, so reinforced in the human world, that we believe that the two extremes of Aristotle's liberality are good, making money and giving some of it back. Not only that, we have sanctified giving back the money we have made legitimately. We have made charity as the emblem of benevolence and, thus, the more money we make, the more benevolent we can be with those we help out of pity for their misery. However, we should start thinking whether that is an act of generosity or another form of *hypocrisy* we, human beings, are reinforcing in the human world. *Do we really care about others or do we want to appear to be benevolent when what drives that "generosity" is coupled to the feeling of guilt that goes with reinforcing greed?* Giving back to a few needy people out of pity does not prevent the suffering of others, which is what generosity is about, especially if reinforcing greed will only increase the number of those who will become poor and will have to be released from their suffering through more charity. However, there is only one way of being benevolent, rather than appearing so by giving money, and that is by realizing the value of generosity through what we do for others to prevent having to release the suffering of some "fortunate" souls. Before that day comes, we must come to the realization that receiving money with one hand and giving back to the "poor" with the other, does not compensate for the amount of suffering that could be prevented worldwide from realizing the value of generosity.

Maybe we should start by trying to *avoid conflating philanthropy with generosity if that help comes from pitying others.* The love of humanity is so important to save our world that we cannot afford to take its name in vain. Likewise, we cannot continue describing the countervalue related to the value of generosity *making money or doing well* when it is nothing but *greed*. Making a profit is not something bad *per se* but absolutely necessary to provide something of value to others. One thing is making profit as a consequence of providing something of value to others and quite another is *the profit motive for providing something of value to others*. But when did greed and its offspring, charity, get sanctified? Well, we have already seen what Aristotle had to say about making money for its own sake, which also affects the political profession, "most take up the 'political' life for the sake of money and greed". However, I doubt he would have said the following if he had lived in our time: "The life of the illiberal is servile, suited to a slave, and sordid, remote from ambition". For Spinoza (1677), greed or "avarice is the immoderate desire or love of riches". Kant (1775–1781), however, was ambivalent about riches "because of what they can provide": "To possess wealth sufficient for any purpose is in itself pleasant". That changed with Hegel. To quote from a conference delivered by Angel Ros (1998), my high school philosophy teacher,

> Hegel proclaims the ethnicity of lucrative activity and the honesty in the use and exchange of wealth, which he himself grounded on the religious consciousness of Protestantism, contrasting it with the false ethics of poverty of Catholicism.
>
> (Ros 1998[2002]: 107)

Furthermore, Hegel ([1820] 2001) sanctified the people who made money in their trade and gave back to charity "out of sympathy".

> In the corporation the assistance received by poverty loses its lawless character, and the humiliation wrongly associated with it.

> The opulent, by performing their duty to their associates, lose their pride, and cease to stir up envy in others. Integrity receives its due honour and recognition.
>
> (Hegel [1820] 2001: 193)

Greed is no longer despicable, because someone has to release the suffering of the "pauper class" and do something about "their immediate need, dislike of work, bad disposition, and other vices". This is the same feeling of superiority we saw coupled to nationalism but now towards others, which also has a name, *condescendence*. But this is far from springing from the love of humanity that moves people to help others because it has its source in misanthropy, *the contempt for other people*.

Since countervalues are mutually related, let us continue with *selfishness* after greed. Well, in Chapter 3 we already saw that Aristotle (1915) defines ethics as the pursuit of happiness which, of course, implies that countervalue. What I did not say then, however, is that the true pursuer of happiness does not pursue any sort of happiness but only what is good and noble in life which is why "the good man loves most this part of him",

> [as] in respect of the noble and good he is loving himself most; for he is compassing these for himself as being noblest. He is therefore a lover of good, not a lover of self. For, if he does love himself, it is only because he is good.
>
> (Aristotle 1915: 1212)

Therefore, the virtuous person is selfish but not in the ordinary sense in which most people are but in the excellent or noble way. Indeed, good selfishness is that of great or heroic people who stand out from the rest as "the dignified and the magnificent" and "since magnanimity seem to accompany *all* the other virtues", they are "perfectly virtuous" (Aristotle 1915: 1207). So that selfishness is very rare and since there are also intellectual virtues besides ethical virtues, those exceptional human beings are practically *and* theoretically wise. Regardless of how uncommon those people are, selfishness is always selfishness and arrogance is selfishness coupled to pride-self-love which despises other human beings as being inferior. Well, how different is this from Spinoza's (1677) understanding of selfishness which impregnates the whole of nature. "Everything in so far as it is in itself endeavours to persist in its own being". This selfishness anticipated Schopenhauer's Will that manifests in multiple individuals. Though Hume (1739) discarded selfishness to explain morality, he did claim that it was "the original motive to the *establishment* of justice". And we have already seen how Smith gave priority to self-care in his hierarchy of proximity in Chapter 8. Finally, Kant identified selfishness with "the inclination" and argued that the moral motive suggested an altruistic form of selfishness.

Schopenhauer (1840), on the other hand, also saw selfishness as an "anti-moral motive". "The chief and fundamental incentive in a human being, as in an animal, is *egoism*, i.e. the urge to existence and well-being" (Schopenhauer [1840] 1903: 151). However, what distinguishes the animal from the human being is "self-interest", which "denotes Egoism, so far as the latter is guided by reason, which enables it, by means of reflection, to prosecute its purposes systematically" (Schopenhauer 1840 [1903]). In other words, selfishness moves human beings to pursue purposes. And it is, in fact, *those things that we do for others to benefit ourselves what reinforces selfishness* in the human world. However, whoever wants to counter that countervalue, will suffer the consequences. "Everything that opposes the strivings of his Egoism awakens his dislike, his anger, his hate: this is the mortal enemy".

> Egoism is a huge giant overtopping the world. If each person were allowed to choose between his own destruction and that of the rest of mankind, I need not say what the decision would be in most cases. Thus, it is that every human unit makes himself the centre of the world, which he views exclusively from that standpoint.
>
> (Schopenhauer 1840 [1903]: 151–152)

Schopenhauer was right, selfishness, like every other form of misanthropy, is coupled to the emotion of anger and, indeed, it could go as far as destroying our world, though no so much from pressing a button detonating an atomic bomb as, he loves himself too much to do that, but from gradually destroying it through what they do for others to benefit themselves. Indeed, there is *nothing more powerful to destroy our world than a combination of greed and selfishness* coupled to pride-self-love and the absolute indifference towards others that comes from these two forms of misanthropy taken together. But to counteract or suppress egoism, as Schopehauer argued, humans do not have to fight it with compassion, which is an emotion coupled to the love of humanity, but with a value instead. In fact, that compassion, if I may say, could be confused with the charity that springs from greed. But what human systems need to realize in the human world is something that goes together with generosity instead. To finish with selfishness, the philosopher of compassion made an insightful classification between the vices derived from egoism, in which he included greed, and those derived from hostility that already suggests an indifference towards others. The difference between greed and selfishness is that the former is reinforced by people who do something for others to enrich themselves and the latter do so to gain something for themselves, which can include other things besides wealth or money.

Our next value to which selfishness or egoism is related is one which no other philosopher except for Scheler (1913–1916), and later Hartmann (1926), had mentioned before and he originally conceived it as a theoretical principle to distinguish between the "essential forms of human groups". In particular, to set the collective person apart from "the society" in which "[t]here is no true *solidarity*" (my emphasis), "it is a non-representable solidarity characteristic of self-interested individuals".

Whereas,

> in the collective person every individual *and* the collective person are *self*-responsible (= responsible for oneself), and at the same time every individual is also coresponsible for the collective person (and for every individual "in" it), just as the collective person is coresponsible for *each* of its members.
>
> (Scheler [1913–1916] 1973: 533–534)

Since values are realized by human systems, that is, by humans working together for a purpose, *the value of solidarity is realized by doing something together to help others*, given that the values of solidarity and generosity are mutually related. However, Scheler's concept of solidarity was inward looking because that mutual help between the members of the collective person is contrasted with "the society" in which people only cooperate for self-interest. Unfortunately, he was trying to accommodate the self-interest of individuals within the collective person with the "coresponsibility" of every member for each other. Someone might have already noticed that, besides trying to integrate selfishness within a human system, Scheler could not see beyond the artificial division between the individual and society, using formulas such as "one for all" and "all for one" to describe "true solidarity". However, we cannot combine a value and a countervalue together to form a hybrid value as it is logically impossible to love solidarity and selfniness simultaneouly. Furthermore, if values are realized through what human systems do for others, solidarity is something that they realize rather than something they share together. In fact, what people who realize solidarity share is the purpose of the human system in which they do something for others. Therefore, it is not by helping each other that they realize the value of solidarity but by helping others besides themselves through what the human system does to help them. Indeed, people could work together for themselves rather than for others and would not realize the value of solidarity but reinforce the countervalue of selfishness.

And human systems who realize the value of solidarity helping others do not do it out of pity but out of the love for others, that is, without condescendence. This is also implied in Scheler's understanding of solidarity which is not conflated with sympathy but implies it. This is suggested by the distinction he makes in a latter work on the *Essence and Forms of Sympathy* ([1923] 1950) between different types of co-experiences in which love tops sympathy. What he could not avoid, however, is treating solidarity as a co-feeling experienced by the people who belong to the collective person. Furthermore, the tone in which he discussed the collective person to distinguish it from the civil society – given what he had in mind, a society in which its individuals sign contracts for which they were accountable – made him use a language related to the juridical sphere of the human world. This could explain why Hartmann ([1926] 2011) discussed solidarity in relation to justice as the "consciousness of such coresponsibility" as a member of a community which is "the foundation of his legal and civil solidarity with the multitude of equals". However, he went beyond Scheler arguing that it was also related to brotherly love that extends beyond the boundaries of a nation-state.

> For although only one who is near is the object of sympathy, yet potentially every human being is a neighbour – and this is no mere possibility, for there are many ways of coming into contact. But, above all, through brotherly love the fact of being jointly responsible grows into universal participation in the fate, the sufferings and actions of the rest of the world.
>
> (Hartmann [1926] 2011: 497)

This is more like the love of humanity that transcends Smith's hierarchy of proximity, even though Hartmann's understanding of solidarity still mentions sympathy and, unlike Scheler, it is something centred on the individual human being. He is aware that, unlike the juridical solidarity, it "correspond[s] to a different sort of union between person and person". And yet, he could not see beyond the individual who truly cares about others. "It does not bear the whole social structure on its back, it resists any fixed form. Yet it by no means spends itself whole in the strong sentiments which bind us to those who are near and dear to us". Without *doing something together* to help others, that is, without human systems, the value of solidarity would not be realized in our world. But this does not imply that we can only realize this value at a certain scale or by human systems of a certain dimension. Instead, the fourth fundamental law of ethics applies to all scales of the human world, from the subnational to the supracivilizational levels.

But people could still be doing something together to help others not out of pity but out of obligation or a sense of duty towards others. Yet, they would not be realizing the true value of solidarity because they would not be doing it out of the love for others. So there is something missing that is pointing us towards another value that suggests that the realization of the value of humanity would be incomplete without it, namely, the *value of friendship*. All the moral philosophers who have preceded us have mentioned it and even some who I have not mentioned, such as Epicurus, for whom "Of all the things which wisdom provides to make us entirely happy, much the greatest is the possession of friendship". Since friendship is a value, I will have to question those philosophers, such as Aristotle and Kant, who distinguished between different forms of friendship, according to whether we have friends for pleasure or utility or because we like people who are good, like us, as Aristotle believed, that is, "everyone loves what is good for him" and the highest form of friendship is between good people, which Kant called "*moral friendship*". However, though both philosophers agreed that it depended on reciprocity, since it had to be mutual, for Kant it was based on "reciprocal respect", not affection. Again, he was stepping into the juridical sphere in which we have to respect the law which, in his case, meant respecting the moral law or categorical imperative dictated by reason. Another difference is that for Aristotle friendship took place between equal *and* unequal people, in which one was superior to the other but the inferior provided something "proportional to the superiority of the other". Similarly, for Kant, friendship compensated for the enmity between people to which it was contrasted.

Well, friendship is neither a desire we have to satisfy or a pleasure derived from other people's company nor is linked to utility because this value is not related to the countervalue of selfishness, but to the value of solidarity, and, as such, human system *help others who they treat as friends* out of the love for others. Indeed, to use one of Kant's (1785) formulations of the categorical imperative: human systems act regularly in such a way that they always treat humans, "never simply as a means but always at the same time as an end". This means that human systems who realize the value of friendship not only do not utilize others for their own selfish interest, but also *treat themselves as friends* out of the same love for others. In other words, if they only treat as friends the people who they serve and do otherwise among themselves, then they cannot realize the value of friendship. Kant was well aware that we are not angels and that misanthropy will never be eradicated completely, so he preferred to think about friendship as an ideal that was itself unrealizable. All that we could hope for was to achieve the maximal degree of friendship between us, "through the same love and mutual respect". "Man is a being destined to live in society (although also unsociable, however)". Well, it is difficult to treat others as friends if we believe that human beings are antagonistic by nature, as Kant (1784) claimed in the *Idea for a Universal History with a Cosmopolitan Purpose*. "The *unsocial sociability* of human beings, i.e. their propensity to enter into society, which, however, is combined with a thoroughgoing resistance that constantly threatens to break up this society" (Kant [1784] 2000: 46), Again, we cannot integrate "hatred of mankind" with the love of humanity, we either love humanity or have contempt for it, but we cannot have both at the same time.

As I said earlier, there are two forms of misanthropy related to the indifference towards humanity that we have already seen, namely, greed and selfishness, and the other two remaining forms of contempt for humanity are related to hate, namely, *cruelty* and *resentment*. Though they are mutually related, what interests us now is to clearly distinguish the countervalue related to friendship which is cruelty, not enmity, as Kant suggested, because, again, values and countervalues are not opposites. Aristotle would relate both countervalues to Plato's irascible or spirited part of the soul whose virtue was *courage* or valour, which avoided the excess of the lack of fear, namely, temerity or recklessness. As "courage has to do with fears, so that moderate fears increase courage", but also the excess of fears that results in the contrary vice, cowardless. And, though Plato also related the irascible part to *gentleness* as the opposite of temerity, Aristotle would treat it separately because, according to him, it was "a mean state" between the lack and excess anger. However, he contrasted the irascibility, "a vice of the passionate part, through which men are easily stirred to anger", with cowardice, "vice of the passionate part, through which men are scared by apprehensions, especially such as relate to death". Indeed, the reinforcement of the hatred of humanity by human beings and human systems is coupled toh the emotion of anger, since fear is coupled to the love of life, as we will see in the last chapter.

In the case of Spinoza (1677), cruelty is defined in a way that comes close to resentment, "the desire whereby any one is incited to work evil to one whom we love or whom we pity" (Spinoza [1677] 1970: 137), because the *harm to others* is not done indiscriminately as with cruelty, which is reinforced by people who do not treat others as friends. Therefore, *cruelty can be defined as the contempt for everyone whom someone hates and resentment as the contempt for those people whom someone hates*. However, though these two countervalues are different, they are more closely related than we would expect. Let us think about vengeance, where would we place it? It is a form of cruelty, surely. But, equally, the desire for revenge is a form of resentment. Indeed, it is a form of resentment that can end in cruelty when the people whom someone hates are harmed by him. The hatred of humanity of the people who reinforce cruelty in the human world, through pursuing purposes that harm others, does not need much discussion as a contervalue related to the value of friendship. It is clear that people who do not treat others as friends, not only do not care about them but want to harm them and will continue reinforcing cruelty whenever they have the chance to do it until they stop hating them. Well, maybe I should say that cruelty reinforces cruelty, as captured by Spinoza's definition of vengeance, "the desire by which we are incited through reciprocated hatred to work evil to him who has worked evil to us" (Spinoza [1677] 1970). And that not all forms of cruelty are so extreme, such as derision or mockery, which can also hurt others.

Now that we have untangled cruelty from resentment, it is time to discuss the last countervalue that will bring us face to face with the last value completing the last fundamental law of ethics. Fortunately, we have a moral philosopher that can aid us here who treated it in a work under the same name, *Ressentiment* ([1912] 1998), namely, Scheler. This book was a reaction to Nietzsche's thesis that Christian ethics was born out of resentment towards what preceded it, "noble morality". Essentially, resentment is a repeated "emotional response reaction against someone else" and whose nature is negative, as it entails "hostility", and the word "rencor" is what "comes closest to the essential meaning of the term", according to Scheler. Well, as to Nietzsche's thesis in the *On the Genealogy of Morals* ([1887] 1996) was that "Christian love" was the result of a "slave revolt in morality" that "begins when *ressentiment* itself becomes creative and produces values". As usual, and contrary to the relationship between values and countervalues, he wants to derive a value from its antithesis. But to cut a story short, that "slave morality" was born "out of the stem of that tree of revenge and hatred, of Jewish hatred" as it could not come out of itself, "in order to arise, slave morality always needs a hostile external world". In addition, he sees forgiveness not as a value (I am anticipating myself) but as submission, hence "slave morality".

> Weakness is to be made a *merit* [...] and impotence, inability to retaliate, is to become "goodness"; timorous lowliness becomes "humility"; submission to those whom one hates is "obedience" (obedience toward one of whom they say that he decrees this submission – they call him God). The inoffensiveness of the weak, even the cowardice in which he is rich, his unavoidable obligation to wait at the door acquires a good name, as "patience", it is also called virtue; the inability to avenge oneself is supposed to be a voluntary renunciation of revenge, sometimes it is even called forgiveness" ("for they know not what they do – we alone know what they do"!).
> (*On the Genealogy of Morals*, Part I, Section 14)

As we see from the last sentence in brackets, forgiveness is coupled to pity and condescendence in the sense that we have to pity those who do not know any better and forgive them for not knowing that what they are doing is bad. And he ends that passage with sarcasm: "They also speak of 'love for one's enemies,' – and they sweat while doing so". Indeed, Nietzsche cannot hide his contempt for the antithesis of "noble morality".

Talking about coupling and continuing with Scheler ([1912] 1994), he tells us that resentment is related to certain "emotions and affects", such as "revenge" and "hatred" but also "malice" and "envy", claiming that the first is the "most important source of ressentiment" and is coupled to "anger and rage". And he is right in claiming the relation between resentment and the emotion of anger, but not in claiming that "Another source of ressentiment lies in envy", however, unless that envy is related to peoples whom someone hates whom have something they want. Unfortunately, while I am writing the book, there are two ongoing wars in different parts of the world that are cases in point, namely, the Russo-Ukrainian war and Israel-Hamas war. Of course, it goes without saying that it reinforces the countervalue of resentment in our world and the contempt for other peoples.

At last, I can finish this chapter with the *value of forgiveness* which is realized out of the same love for others that inspires the realization of the values of generosity, solidarity, and friendship. I do not know of any philosopher who has treated this value aside from a few remarks from Kant (2002) in an essay "Against Moses Mendelssohn" (1793) and in his *Perpetual Peace* (1795). According to him, in the same way as a people decided voluntary to submit themselves to the "public law" that reason dictates and constitute a *civil state*, "the distress produced by the constant wars in which the states try to subjugate or engulf each other must finally lend them, even against their will, to enter into a *cosmopolitan* constitution" ([1793] 2002: 170). Indeed, the perpetual peace among peoples depended theoretically on *establishing* a "state of peace" in which "No state shall violently interfere with the constitution and administration of another" because, unless all the neighbours are subjected to the same *cosmopolitan law*, "one [nation] is at liberty to challenge another and treat [it] as an enemy" ([1795] 2002: 198). Is this not that what we tried to do after the Second World War with the ratification of the Charter of the United Nations? Kant wanted to ground morality on duty, that is, on the obligation to respect the moral law dictated by reason and, thus, believed that the peoples from all over the world would eventually submit themselves to the same law to come out of the state of nature in which they coexisted at the international level. Presumably, when people come out of the state of nature they agree to stop using the force against each other and forgive all the mutual harm that they have done to each other until then. Of course, it does not work that way because the only thing that can stop the hostility between peoples is to stop hating each other which is the only possible foundation for a perpetual peace. And for that to happen, they have to forgive each other and assume the guilt that is coupled to reinforcing the suffering of others.

However, there is an even more threatening and preventable suffering against ourselves that can end the miracle of life that has been granted to our species and that demands an even greater form of redemption that comes from acknowledging that we, humas, are guilty of the ecological holocaust we are causing through what we do for ourselves, and that only we, humans, can save the destruction of our world. And for that to happen, we have to realize the value of forgiveness repairing the damage we have caused to our world and our planet out of contempt for humanity and life on Earth. That redemption is not only possible but *the only possible one that will be granted to us before the miracle of life is over for us*. And that final redemption, if we make it, will only come from a change of direction, from a new horizon for the human species on this planet and in the universe, if we prevail. And we already know how our new world must look like, it must be shaped by universal values or it will not be. So we have to start rebuilding our human world realizing the "new ethics" dictated

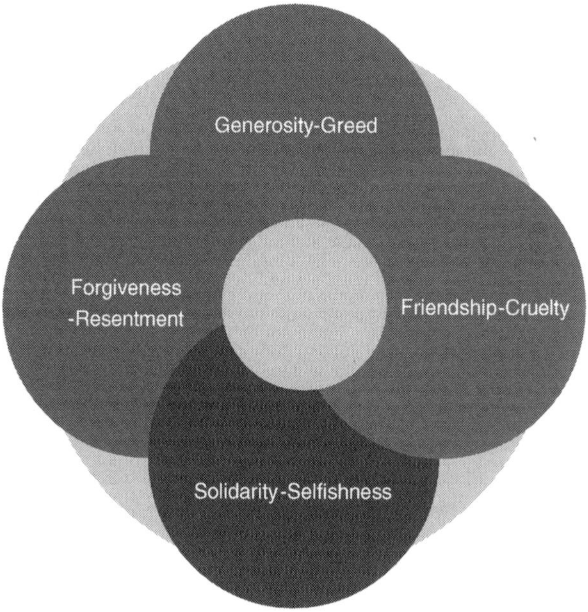

Figure 9.1 The Fourth Fundamental Law of Ethics: Humanity-Misanthropy.

(not by ourselves) by the fundamental laws of the human world, and through the purposes prescribed by what we love and will continue cherishing out of our caring about others and for as long as we may endure.

We can conclude this chapter with the set of values and countervalues that constitute the fourth fundamental law of ethics (Figure 9.1).

References

Aristotle (1910). *Nicomachean Ethics*, trans. G. H. Lewes. London: George Routledge.

Aristotle (1915). *The Works of Aristotle*, trans. W. D. T. Ross, Oxford: Clarendon Press.

Hartmann, N. (2011). *Ética*. Madrid: Ediciones Encuentro.

Hegel, G. W. F. (2001). *The Philosophy of Right*. Kitchener, Ontario: Batoche Books.

Hume, D. (2000). A *Treatise of Human Nature*. Oxford: Oxford University Press.

Kant, I. (1988) *Lecciones de Ética*. Barcelona: Editorial Crítica.

Kant, I. (2000). *Filosofía de la Historia*. Madrid: Fondo de Cultura Económica.

Kant, I. (2001). *Fundamentación de la Metafísica de las Costumbres*. Madrid: Espasa Calpe.

Kant, I. (2002). *Història i Política*. Barcelona: Edicions 62.

Nietzscher, F. (1996). *On the Genealogy of Morals*. Oxford: Oxford World's Classics.

Ros, A. (1998). ¿Qué nos falta aún para ser "Estado"?, in *Escrits* (2002). Girona: El Punt and Hermes Comunicacions. https://dialnet.unirioja.es/congreso/edicion/1556

Scheler, M. (1950). *Esencia y Formas de la Simpatía*. Buenos Aires: Losada.

Scheler, M. (1973). *Formalism in Ethics and Non-Formal Ethics of Values*: *A New Attempt toward the Foundation of an Ethical Personalism*. Evanston: Northwestern University Press.

Scheler, M. (1994). *Ressentiment*. Milwaukee: Marquette University Press.

Schopenhauer, A. (1903). *The Basis of Morality*. London: Wan Sonnenschein.

Smith, A. (1982). *The Theory of Moral Sentiments*. Indiana: Liberty Fund.

Spinoza, B. (1970). *Ethics*, trans, A. Boyle. London: Heron Books.

10 Emotions and Feelings

We cannot exclude the treatment of emotions and feelings from ethics given that it studies the laws that apply to the moral sphere of the human world *in relation to human systems* who are the shapers or demiurges of the human world. Human systems and human beings are equality constituted since they share the same anatomy. Moreover, this means that they both share the constitution of sentient beings which are cybernetic systems in a stable equilibrium with their environment. In other words, *human systems share the anatomy of cybernetic systems*. This understanding of humans is at odds with the traditional concept of human beings as constituted by mind-body dualism and society as a collection of individuals. In the same way as we have to do away with those two prejudices, we also have to question the understanding of animals inherited from moral philosophers which is derived from those same dualisms. However, as Newton said, we can only see furtherby standing on the shoulders of giants and, thus, we are going to learn as much as we can from their insights into emotions and feelings. Philosophers saw from the very beginning that emotions, indistinctively called passions or affections, could not be excluded from the field of ethics, contrary to what Kant was trying to do by grounding ethics on reason and freeing the person from the influence of any external motive or inclination which included emotions. However, as we saw with Scheler and Hartmann, they could not separate ethics from our feelings or emotional acts through which we experience values. Indeed, the realization of values and reinforcement of countervalues cannot be separated from the emotional underpinning of human beings and human systems. In short, *emotions and feelings are morally relevant because they are inseparable from what humans pursue in life*.

DOI: 10.4324/9781003649496-12

For starters, we have to be able distinguish emotions from feelings which moral philosophers have usually bundled together. Emotions are tied to the same anatomy that we share with animals and, as such, are shared with animals, whereas feelings are something unique to humans which cannot be found in animals because they are differently constituted. So we share the same emotions that animals experience but feelings are exclusively felt by humans. In terms of what emotions we experience, some philosophers have been more generous than others in granting them to animals or denying them what comes from their cybernetic constitution. The most generous have been Hume and Schopenhauer who understood that we cannot deny to animals what we share with them. The more restrictive were Spinoza and Kant who were less kind to their nature because they lacked the thinking substance that makes us unique in the universe. Aristotle did not go that far because he inherited from Plato the appetitive part of the soul that brings us closer to "beasts" with which we share the sensitive soul. As to feelings, since they conflated them with emotions, most were treated as virtues-vices or good-bad affections that we either approve or disapprove. And this is where the *moral conscience* – not to be confused with Descartes's concept of the subject or consciousness – comes into the picture. Indeed, conscience is what is unique to humans and what *makes us care about what we do or do not do for others.*. This is what distinguishes humans from animals, we pursue things that matter to us and feel responsible for them and this is why feelings accompany the realization of values or reinforcement of countervalues in our world through what we do for others regularly.

So let us start with emotions that accompany what humans pursue in life and thus are related to values and countervalues, but in a more fundamental way because they are what connect us to the world around us. Since we share the cybernetic constitution of animals, we are also functionally related to our environment through a feedback mechanism in which stimuli from the environment have an emotional response from the organism and that reaction depends not only on the stimuli received but also on their familiarity or exposure to that environment. However, some emotional reactions are inherited from the species or population to which it is genetically related through its evolutionary history. We do not need to dig deeper into the laws that apply to the biological and cybernetic kingdoms, all we need to know is that the species that live today in our planet have survived as a result of a gradual process of natural selection. Furthermore, and this is not explained by natural selection but by Ross Ashby's theory, cybernetic systems maintain a stable equilibrim with their environment as a result of a constant process of information erasure that explains adaptive behaviour. Therefore, emotional reactions have survival or adaptive value and without them animals would be less successful at adjusting to their recurring environment that changes during their lifetime. Without being capable of changing behaviour, they would depend exclusively for their adaptation on the fixed behaviour genetically inherited from species that has evolved through thousands and millions of years of natural selection.

To my knowledge, Darwin may have been the first naturalist to devote a full book to the study of emotions, *The Expression of the Emotions in Man and Animals* (1872), rather than including it as an additional chapter in *The Descent of Man* (1871).

As a natural scientist, Darwin wanted to detect emotions through observable expressions though some were not easy to detect empirically.

> A man may have his mind felted with blackness hatred or suspicion, or be corroded with envy or jealousy; but as their feelings do not at once lead to action, and as they commonly last for some time, they are not shown by any outward sign.
>
> (1872: 79)

Anyway, he was sure that "Hatred and Anger" and "Fear-Horror" were emotions that deserved a chapter as there are had clear external signs of their existence in the animal kingdom because, as Schopenhauer noticed, animals do not hide what affects them which is something that only humans are capable of, namely, dissimulation. Well, the first physiologist to discover that those two fundamental emotions were paired together was Walter Cannon ([1915] 1927), the so-called *fight or flight response* in animals, which he linked neuroanatomically to the sympathetic division of the nervous system.

> Fear, rage, and pain, and the pangs of hunger are all primitive experiences which human beings share with lower animals" [and] "all body changes which occur in the intense emotional states – such as fear and fury – occur as result of activity in this division".
>
> (Cannon [1915] 1927: preface, 268)

Likewise, the pairing together of the other set of fundamental emotions to which Darwin alluded in referring to one of its members, envy, not quite sure of what to make of it, had already been suggested by Plato, *envy-compassion*, in *Gorgias* (380 BC), a dialogue from the Socratic period. In fact, Socrates, according to Xenophon's *Memorabilia* (1899), was the first to define envy.

> Considering what *Envy* was, he decided it to be a certain uneasiness, not such as arises, however, at the ill success of friends, nor such as is felt at the good success of enemies, but those only, he said, were envious who were annoyed at the good success of their friends.
>
> (Xenophon 1899: 128)

And maybe the first moral philosopher to see that emotions were related to pain and pleasure was Aristotle (1910). "By Emotions, I mean such as lust, anger, fear, confidence, envy, joy, friendship, hatred, longing, emulation, compassion, in short all such as are followed by pleasure or pain" (Aristotle 1910: 43). Other times (1915) he reduces the list to "anger, fear, hate, desire, envy, compassion" and, similarly to Socrates, "envy is pain felt at deserved good fortune, while the feeling of the man who rejoices at misfortunes has itself no name, but such a man shows his nature by rejoicing over undeserved ill fortune" (Aristotle 1915: 1185, 1233). Well, regarding the latter shorter list, except for hate and desire, which are related to need, all those emotions which he mentions, and the concomitant pleasure and plain that come with them, are also found in animals. Of course, we have to define them in a simpler manner because animals are only aware of what they perceive and cannot guess the fortune and misfortune of others in the case of compassion and envy. Sentient beings experience those emotions when they are *satisfied or unsatisfied with the pleasure or pain of others because they like or do not like seeing them in that situation*. And in the case of fear or anger, sentient beings experience those emotions when they *avoid or confront an impending threat or danger that can harm them because they do not like to suffer.*

Despite the fundamental nature of both sets of emotions, moral philosophers were not sure whether they all applied to animals because it implied granting them a high degree of intelligence. Hume (1739), for instance, although accepting that their mind works like ours and that "*sympathy*, or the communication of passions, takes place among animals, no less than among men", thought that "Envy and malice", which are "very remarkable in animals", "are perhaps more common than pity". But it never occurred to him that envy and compassion always go together. Kant (1775–1781) was of a similar opinion but more radical, however, "humanity consists in sympathy and pity, since these differentiate man from the beasts". Spinoza (1677), on the other hand, neither granted pity to animals nor had compassion for them. "Pity is pain accompanied by the idea of an ill which happened to another whom we imagine similar to ourselves", hence, it does not apply to animals. Fortunately, Schopenhauer ([1840] 1903), [1851] 2009) was more kind to sentient beings but, from his writings, it seems that compassion was something exclusively human since "compassion itself is an undeniable fact of human consciousness". In fact, this emotion was so fundamental that he regarded it as "the real basis of all *voluntary* justice and all *genuine* loving-kindness". Likewise, he saw envy as "eminently human" and was well aware that we cannot have one emotion without the other.

> It presents itself, as already said, just where sympathy should appear; envy, on the contrary, only there where no occasion for the latter exists, but rather for its opposite, and as this opposite it arises in the human breast, and is therefore in so far a human emotion; I fear, indeed, that no one will be found entirely free from it.
>
> ([1851] 2009: 235)

Thus, fundamental emotions are paired together and come together with pleasure and pain (Table 10.1). As to the exact relationship with these opposite experiences, it is clear that compassion and envy are related to both but in a reverse way given that one can experiences pleasure where another experiences pain. In the case of fear and anger,

Table 10.1 Relationship between Emotions and Experiences

Emotions	*Experiences*
Compassion	Pleasure-Pain
Envy	Pain-Pleasure
Fear	Pain
Anger	Pleasure

however, it is not that easy to see that they are opposite emotions because fearing a threat or danger, which produces pain, does not seem the opposite of fighting back that threat or danger, which does not produce pleasure *per se,* at least not in the way we usually understand pleasure. And yet, fear and anger are opposite emotional reactions to the same stimuli which the sentient being experiences. Well, this apparent contradiction in terms of the concomitant experiences that accompany these emotions is only apparent because pleasure and pain are not opposite in the ordinary sense. Sentient beings experience pain and pleasure but to ease or remove a pain also produces pleasure. Perhaps the first to notice the latter was Schopenhauer ([1818] 1909) who said that, of the two, only pain was positive.

> All satisfaction, or what is commonly called happiness, is always really and essentially only negative, and never positive. It is not an original gratification coming to us of itself, but must always be the satisfaction of a wish. The wish, i.e., some want, is the condition which precedes every pleasure. But with the satisfaction the wish and therefore the pleasure cease. Thus the satisfaction or the pleasing can never be more than the deliverance from a pain, from a want.
>
> (Schopenhauer [1818] 1909: 411–412)

If so, the release of pain is equal to the experience of pleasure and, therefore, anger is the lack of pain that comes from fear because both emotions are paired together.

Before we move on to show how these two sets of opposite emotions are related to values and contervalues, is it important to bear in mind that another impediment to our understanding of emotions and feelings is the artificial division between the individual and society that implies that all experiences are private and that we can only guess what affects others by imagining ourselves going through the same. So we cannot experience the same exact pain but something similar because we are also sentient beings who suffer similarly. However, this way of thinking discounts the fact that cybernetic systems and human systems can *share the same emotion if they share the same environment* with other sentient beings, even if they do not all experience the same in front of the same situation. In other words, we have to start thinking in terms of systems because entangled beings are what the universe is made up of, that is, beings that share a spatio-temporal structure. Indeed, human systems can experience the same emotion when facing the same situation because they share the same environment and may have the same emotional reaction to that situation. Again, this does not imply that all the human beings that face the same situation will have the same emotional reaction to it.

As we know, human beings and human systems can reinforce countervalues but only human systems realize values in the human world. Therefore, human systems are the true shapers or demiurges of the moral sphere of the human world and given that they share the same anatomy with cybernetic systems, the realization of values and reinforcement of countervalues will be coupled to the same fundamental emotions that we share with sentient beings (Table 10.2). I have already mentioned some of those couplings in relation to the values of fellowship and humanity because they were useful to introduce the corresponding countervalues in order to distinguish them properly from the values to which they were related. Well, now is time to unveil the remaining couplings between the fundamental emotions and the fundamental values-countervalues. To be more precise, we should replace "coupled to" with "correlated with" in order to distinguish it from the couplings between emotions and pleasant or painful experiences. We should speak of a correlation of values-countervalues with emotions given that they come together but can be isolated from each other, whereas in the case of the relationship between emotions and pleasure-pain that is impossible. We can experience compassion or envy, for instance, without necessarily realizing the value of humanity and reinforcing the countervalue of nationalism with which they are correlated. However, the experience of those emotions necessarily implies the pleasure or pain coupled to them.

Contrary to what moral philosophers have thought, there are not good and bad emotions. Envy and anger, for instance, are not necessarily bad as they are both correlated with values *and* countervalues. Aristotle avoided doing so by distinguishing affections related to vices

Table 10.2 Correlation between Values-Countervalues and Emotions

	Compassion	**Envy**	**Fear**	**Anger**
Values	Humanity	Knowing	Life	Fellowship
Contervalues	Suffering	Nationalism	Ignorance	Misanthropy

from virtues as the mean between two extremes. However, he did say that "Envy then tends to injustice (for the acts arising from it affect another)" (1915: 1233) and, conversely, that "there are occasions on which anger is a duty" (1910: 63). In the case of Spinoza (1677), what brings us pleasure is good and what brings us pain is bad because the former is a "transition from a less state of perfection to a greater" and the latter, a "transition of a greater state of perfection to a lesser". The fact that Hume (1739) treated envy together with malice implies his predilection for compassion against "a kind of pity reverst". However, for him pity, and malice, were not fundamental emotions, but "secondary ones, arising from original affections", which explains why it was uncommon to find pity in animals. He also treats anger together with benevolence but to show that they produce opposite affections as to those experienced by others, "joy from grief of others, and grief from their joy" and to reconcile this apparent contractions he suggests that "there is always a mixture of love or tenderness with pity, and of hatred or anger with malice" (Hume [1739] 2000: 245). Again, he implied that anger is necessarily a bad emotion. Likewise, Smith (1759) was of the same opinion. In making an analogy with music and contrasting "the modulations of grief and joy", he wrote that "The voice of anger, on the contrary, and all of the passions which are akin to it, is harsh and discordant". "Hatred and anger are the greatest poison to the happiness of a good mind" (Smith [1759] 1982: 37). But not everyone was against envy *per se*. Kant (1775–1781) distinguishes between the envy that came from comparing ourselves to the Idea of perfection or to others. "The first of these methods is sound; the second is not, and it frequently even leads to results diametrically opposed to the first" (Kant [1775–1781] 1988: 260). Finally, for Schopenhauer envy and anger were vices derived from hostility.

Maybe I should say something about the correlation between envy and fear and the values of knowing and the countervalue of ignorance respectively. Well, Kant was on to something when claiming that there are two ways of comparing ourselves to others, distinguishing between a jealousy from finding that others a better than us, which "tries to depreciate it so that his own good points stand out", and "The other species of the genus jealousy, which make us try to add to our good points so as to compare well with another, may be called emulating jealousy" and "is much the less frequent of the two" (Kant [1775–1781] 1988: 261). This clearly applies to the correlation between envy and the value of knowing in which the people who contribute to realize it can make a contribution to knowledge if they are inspired by the great minds who they have met in life or through their works because we can learn a lot from them. Equally, fear is correlated with the countervalue of ignorance because we fear what we do not know and many entrenched prejudices contribute to that fear. In addition, we stop worrying about certain things that preoccupy us once we know what we need to know and can come to terms with it. As to the correlation between fear and compassion and value of life and the countervalue of suffering respectively, we fear losing the opportunity to live or death, which will terminate that gift that has been granted to us, in the same way as we experience compassion for those sentient and human beings who we see suffering because they do not deserve it.

We are finished with emotions, it is time to turn to feelings which are genuinely human as they are related to our moral conscience which is absent in animals. Animals experience emotions but do not feel those feelings that come from realizing a value or reinforcing a countervalue that are only experienced by humans who realize purposes. We will have to start, then, by explaining what moral conscience is so as not to assume that it is something that judges what we do. According to Hume (1739), "moral distinctions", that is, vice or virtue, are founded on a "moral sense" that either "gives a delight or uneasiness; and 'tis from thence the approbation or disapprobation arises" and, therefore, "all the effects of morality must be deriv'd from the same pain or pleasure" (Hume [1739] 2000: 193). In other words, moral judgements are carried out by our mind based on what the moral sense feels. In the case of Smith (1759), that judgement was carried out by an "impartial spectator", "the man within the breast, the great judge and arbiter of their conduct", based on a "sense of praise-worthiness and blame-worthiness" (Smith 1982: 137). Kant (1775–1781) will deny that moral conscience is related to what we feel and will make it the court of reason. "Conscience is an instinct to pass judgement upon ourselves in accordance with moral laws", "and its judgment is not logical, but judicial" (Kant [1775–1781] 1988: 169). It is the moral imperative dictated by reason that judges whether "the maxims of your action", that is, our rules,can be prescribed universally. Kant will not allow morality to be mixed up with emotions but Schopenhauer ([1840] 1903) will not accept that moral conscience is unrelated to our experience. "The entire imagery – that of a judicial drama – whereby Kant depicts conscience is wholly unessential and in no way peculiar to it" (Schopenhauer [1840] 1903: 112).

According to him, what is essential to conscience is that it is related to our deeds and, therefore, Kant was wrong in believing that it appears *prior* to doing something, because it comes with acts and any time it precedes them what appears is "the recollection of previous cases, in which similar acts after they were committed received its disapproval" (Schopenhauer [1840] 1903: 106). So, yes, conscience is related to our actions but, more generally, to our purposes as it comes with them and cannot be separated from them in the same way as emotions cannot be separated from pleasure-pain. However, the concept that we have inherited from our predecessors is one-sided given that it is always associated with the negative, that is, with letting us know what is wrong, in the sense of what is shameful or blameful because what is praise-worthy depends on the judgement of others. In other words, moral conscience has been reduced to a sense of guilt-shame. Schopenhauer even claimed that true conscience has to be distinguished from the regret or remorse that comes from the fear of the negative consequences rather than because we feel guilty of having done something wrong. If we reduced moral conscience to guilt-shame, we would be leaving out all the other feelings that come from pursuing our purposes which do not imply judgement. Furthermore, that sense of guilt-shame would exclude the guilt related to harming non-human beings as that feeling is correlated with misanthropy, whereas the value of humanity is correlated with compassion, as we have seen. Thus, moral conscience cannot exclude other feelings and emotions correlated with other values and countervalues. Indeed, we feel guilty for the ecological holocaust that our species is causing because we love humanity *and* life, not *because of* the negative consequences to ourselves of what we are doing to the planet.

So what are those *universal feelings* experienced by our moral conscience? Well, since they cannot be decoupled from the fundamental emotions that we share with sentient beings, they are also coupled to pleasing and unpleasing experiences but, unlike emotions, those experiences are related to the pursuit of purposes rather than to our environment. And, as such, they are experiences of a different nature which are not necessarily so momentaneous nor so passionate as emotions can be because they can last much longer and maintain their intensity over time and even after the pursuit of those purposes correlated with them is over. We do not necessarily get rid of them once we have finished pursuing something which is not necessarily bad *per se* if what we have pursued is something that brings pleasing feelings but it could also be otherwise. As with emotions, there are no good and bad feelings but only pleasant and unpleasant feelings. To distinguish them from the pleasure-pain to which emotions are coupled to and taking into account how they differ from them, we should speak of feelings coupled to *pleasure-joy* and *pain-sadness* (Table 10.3). Likewise, we will also have to discover the exact relationships between the different feelings and the exactly opposite experiences that our conscience can feel in relation to the pursuit of purposes. This time I will state them upfront before going about introducing each and every one of those feelings with the help of moral philosophers and I will group them according to the relationship that prevails in each of them.

So let us start with *self-love*, which, as you may have already have noticed, I do not separate from *pride* because people who love themselves cannot help feeling proud of themselves for what they have done, but we can be proud of someone other than ourselves which should not be confused with pride as such because it is related to the admiration for what other people have done and can be akin to the respect for someone. Like Schopenhauer said about envy, I do not think anyone is completely immune from the feeling of self-love no matter how much or how little they love-themselves. Spinoza (1677) contrasted *superbia*, "over-estimation of oneself by reason of self-love", which is related to pride, with *abjection*, "under-estimating oneself more than is fair", which is related to *humility*. Pride produces a pleasing feeling "arising from the fact that man regards himself and his power of acting", whereas humility produces an unpleasant feeling from "his want of power or weakness". Basically, people who feel humility through what they pursue in life underestimate what they are capable of, whereas those who experience pride feel otherwise. Aristotle (1915), who was the first to talk about self-love, related pride to the *magnanimity* "which seems to accompany *all* virtues", but instead of contrasting it with humility, he related it to another set of opposed feelings, namely, *vanity-modesty*.

Table 10.3 Relationship between Feelings and Experiences

Feelings	*Experiences*
Pride-Self-Love and Vanity	Pleasure-Joy
Humility and Modesty	Pain-Sadness
Hope and Respect	Pleasure-Joy
Despair and Guilt-Shame	Pain-Sadness

> [T]he quality consisting in thinking oneself worthy of great goods when not worthy is vanity [...] [and] is held to be the mark of the mean-spirited not to think himself worthy of any thing great though he possesses that for which he would justly be deemed worthy of it.
> (Aristotle 1915: 1233)

These two groups of opposed feelings should be differentiated because someone can feel pride for what they have done and also feel modesty when others acknowledge it because they underestimate what others praise highly. However, we cannot say the same for the feeling of vanity which always goes together with the feeling of pride, as the people who are very happy when others recognize what they have achieved feel very proud of themselves. To continue with *pride-humility*, Hume (1739) considered them "indirect passions", "every thing related to us, which produces pleasure or pain, produces likewise pride or humility", and also related vanity with pride, "This happens when external objects acquire any particular relation to ourselves, and are associated or connected with us" (Hume [1739] 2000: 198). And pride with vanity, "by *pride* I understand that agreeable impression, which arises in the mind, when the view either of our virtue, beauty, riches or power makes us satisfy'd with ourselves" (Hume [1739] 2000: 194). Like Aristotle, Smith (1759) also contrasted pride with vanity and, in having to choose between them, "The proud man is commonly too well contented with himself to think that his character requires any amendment", whereas "The desire of the steem and admiration of other people" can lead us eventually to glory, if we have the right "qualities and talents", and, therefore, "The great secret of education is to direct vanity to proper objects" (Smith [1759] 1982: 258, 259). And, like Hume, he also associates vanity with "riches", "because he feels that they naturally draw upon him the attention of the world". Schopenhauer ([1851] 2009), on the other hand, could not bear vanity, "which courts the applause of the world in order to gain the necessary foundation for a high opinion of one's own worth, whilst pride is based upon a pre-existing conviction of it" (Schopenhauer [1851] 2009: 236). Finally, Kant distinguished "proper self-respect", inseparable from the "moral humility" from comparing ourselves to the "moral law", from "noble pride", as a result of comparing ourselves to others, which we saw earlier in relation to jealousy.

As to the other two groups of opposite feelings, let me start with the last feeling, guilt-shame, since Aristotle did not recognize the others as such but I am sure that he felt them, like all humans. You may remember that he considered shame an emotion. Well, he talks about shame awakening courage saying that it "makes a man face danger that the man would most seem to be brave" (1915: 1230). "'Seems" but may not be so since he may do so because of "fear of disgrace" from "low bad actions". True courage or patriotism, if I may say, does not depend on shame. In addition, though not a virtue, shame is the mean state between being modest and shameless to which it is contrasted "since he who fears this [disgrace] is a good man and has a sense of honour, and he who does not fear it is shameless" (1910: 76). So shame is related to noble morality, if I may use Nietzche's expression, based on courage. Spinoza (1677), though he did not relate it to courage, he did contrast honour or glory, which "is pleasure accompanied by the idea of some action of ours which we imagine others to praise", to shame, which "is pain which follows the deed of which we are ashamed" (Spinoza [1677] 1970: 135, 136). We can see that guilt has entered the picture with the feeling of *repentance*, which "is pain accompanied by the idea of some deed which we think we have done by the free decision of the mind". For Hume and Smith, shame is something that affects the individual but comes from society.

To quote Smith (1759),

> It is made up of shame from the sense of impropriety of past conduct; of grief for the effects of it; of pity for those who suffer by it; and of the dread and terror of punishment from the consciousness of the just-provoked resentment of all rational creatures.
>
> (Smith [1759] 1982: 85)

Again, he also contrasts it with the feeling of glory, "esteem and approbation of all mankind", "applauds himself by sympathy with the approbation of this supposed impartial judge" which we carry within us and describes it as "the conscience of merit, or of deserved reward" (Smith [1759] 1982). However, that glory or honour, which can be traced back to Aristotle's magnanimity, implies pride-self-love, which is still present in Kant's concept of dignity or the absolute value of the person but also in Schopenhauer's understanding of pride as self-worth. What we are lacking is a feeling opposed to guilt-shame which cannot come from praising ourselves for what others deem worthy. Instead, that feeling comes from what people have done for others and it is not self-love or vanity but *respect for the pursuit of something worthwhile*. This is not self-respect or self-esteem for what we have achieved because, in contrast to guilt-shame, respect is a pleasant feeling *experienced by the moral conscience of others*. And this feeling is what Kant was struggling with in the *Groundwork of the Metaphysics of Morals* (1785) where he added a footnote to clarify what he meant by respect, after having introduced the proposition that "*Duty is the necessity of an action from respect for the law*".

> One could accuse me of merely taking refuge behind the word *respect* in an obscure feeling instead of giving a distinct reply to the question through a concept of reason. Yet even if respect is a feeling, it is not one *received* through influence but a feeling self-effected through a concept of reason and hence specifically distinguished from all feelings of the first kind, which may be reduced to inclination or fear.
>
> (Kant [1785] 2001: 64)

We are only missing the last two opposite feelings of the second group, namely, *hope-despair*. No other moral philosopher except for Spinoza (1677) can help us here. Rather than treating them one after the other, as he usually does, he intercalates fear and confidence between hope and despair to say that they involve uncertainty. "Hope is an inconstant joy arisen from the idea of a thing past or future, the event of which we still doubt to some extent". "Fear", in contrast, "is an inconstant sadness arisen from the idea of something past or future of whose event we doubt to some extent" (Spinoza [1677] 1970: 131). Well, that fear he talks about is in fact despair when that sadness is constant. He also says that "fear cannot be without hope nor hope without fear" which, of course, comes from the uncertainty of not knowing for certain if the event we hope for or are anxious about will actually happen or not. However, it does not apply to despair because humans who suffers that feeling are not worried because they have surrendered. "Confidence therefore arises from hope, and despair from fear, when all cause of doubt as to the event of a thing is removed" (Spinoza [1677] 1970: 131–132). Since feelings are related to the pursuing something, I would say that the *feeling of hope comes from pursuing something that we are confident we can achieve and the feeling of despair from having lost that confidence.* Now that we have been acquainted with all the pairs of opposite feelings coupled to pleasure-joy and pain-sadness, it is time to introduce the correlation of feelings with values-countervalues to complete the universal moral hypothesis (Table 10.4).

We have already met some of them in relation to the values of fellowship and humanity, such as the feeling of superiority that comes from reinforcing the countervalue of nationalism and the feeling of guilt from reinforcing the contempt for humanity or misanthropy. I have also alluded to the relationship between misanthropy and nationalism through the emotion of envy which means that the feeling of pride-self-love can result in the feeling of guilt from wars or conflicts between peoples that reinforce nationalism and misanthropy in the human world. Likewise, the harm to other people from reinforcing misanthropy also results in the feeling of despair of the people who suffer that injustice. In the case of the value of knowing, the emotion of envy that we saw that was correlated with it is accompanied by the feeling of modesty because the more we know the less arrogant we become about the little we know. In contrasts, the feeling of vanity is correlated with the contervalue of ignorance because the people who think they know everything are not prepared to listen to and learn from other people who are more knowledgeable and, by doing so, they reinforce ignorance e in the human world. But there is a feeling of hope that is experienced by humans who believe that they can pursue something worthwhile that deserves the respect of others, such as saving our world before the miracle of life is over for us (Table 10.5).

Table 10.4 Correlation between Values-Countervalues and Feelings

Values	*Feelings*	*Countervalues*	*Feelings*
Fellowship	Humility	Nationalism	Pride-Self-Love
Knowing	Modesty	Ignorance	Vanity
Humanity	Respect	Misanthropy	Guilt-Shame
Life	Hope	Suffering	Despair

Table 10.5 Correlation between Values-Countervalues and Emotions-Feelings

	Compassion	*Envy*	*Fear*	*Anger*
Respect	Humanity			
Despair	Suffering			
Modesty		Knowing		
Pride-Self-Love		Nationalism		
Hope			Life	
Vanity			Ignorance	
Humility				Fellowship
Guilt-Shame				Misanthropy

References

Aristotle (1910). *Nicomachean Ethics*, trans. G. H. Lewes. London: George Routledge.

Aristotle (1915). *The Works of Aristotle*, trans. W. D. T. Ross, Oxford: Clarendon Press.

Cannon, W. (1927). *Bodily Changes in Pain, Hunger, Fear and Rage*. New York: D. Appleton.

Darwin, C. (1871). *The Descent of Man, and Selection in Relation to Sex*. London: John Murray.

Darwin, C. (1872). *The Expression of Emotions in Man and Animals*. London: John Murray.

Hume, D. (2000). A *Treatise of Human Nature*. Oxford: Oxford University Press.

Kant, I. (1988). *Lecciones de Ética*. Barcelona: Editorial.

Kant, I. (2001). *Fundamentación de la Metafísica de las Costumbres*. Madrid: Espasa Calpe.

Schopenhauer, A. (1903). *The Basis of Morality*. London: Wan Sonnenschein.

Schopenhauer, A. (1909). *The World as Will and Idea*, Vol. I. London: Kegan Paul, Trench, Trübner.

Schopenhauer, A. (2009). *Parerga y Paralipómena*, 2 Volumes. Madrid: Editorial Trotta.

Smith, A. (1982). *The Theory of Moral Sentiments*. Indiana: Liberty Fund.

Spinoza, B. (1970). *Ethics*, trans, A. Boyle. London: Heron Books.

Xenophon (1899). *Memoralia*, trans. J. S. Watson. Philadelphia: David McKay Publisher.

Made in the USA
Las Vegas, NV
19 September 2026

2534505f-d954-40e6-a763-2cb559d8358eR01